Ben Stacy Jerrik (Ed.)

Sprinter (North County Transit District)

Ben Stacy Jerrik (Ed.)

Sprinter (North County Transit District)

North, County, Transit, District, Rail, Line, Technology

Part Press

Imprint

Publisher:
Part Press is a trademark of
International Book Market Service Ltd., 17 Rue Meldrum, Beau Bassin, 1713-01 Mauritius
Email: info@bookmarketservice.com
Website: www.bookmarketservice.com

Published in 2012

Printed in: U.S.A., U.K., Germany. This book was not produced in Mauritius.

ISBN: 978-613-9-00735-6

Contents

Articles

References

Sprinter_(North_County_Transit_District)

Sprinter	
Overview	
Type	DMU light rail
System	North County Transit District
Locale	North San Diego County, California, United States
Termini	Oceanside Transit Center Escondido Transit Center
Stations	15
Daily ridership	8,308 daily [1]
Operation	
Opened	March 9, 2008[2]
Operator(s)	Veolia Transportation
Technical	
Line length	22 mi (35 km)
Track gauge	4 ft 8 $^1/_2$ in (1435 mm) (standard gauge)

The **Sprinter** is a diesel operated passenger rail [3] [4] line operating between Oceanside and Escondido, California, United States. The service uses the pre-existing 22 miles (35 km) Escondido Branch trackage of the San Diego Northern Railroad. Station platforms were constructed for the line's fifteen stations [5] serving the cities of Oceanside (western terminus), Vista, San Marcos, and Escondido (eastern terminus). The line provides service to Palomar College and California State University, San Marcos.

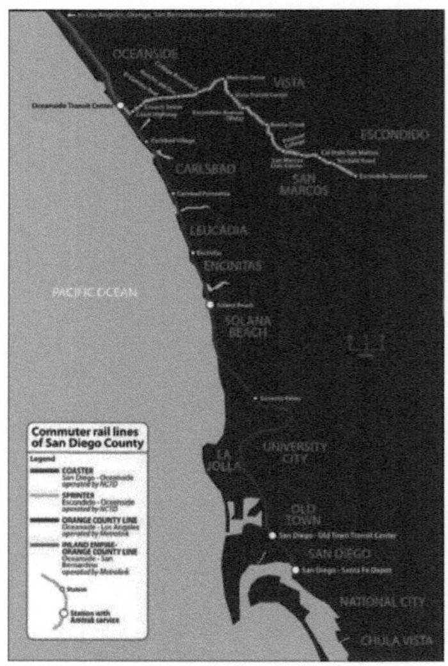

Sprinter map (with other lines included)

The Sprinter is operated by the North County Transit District of Oceanside, the area's public transit agency. The agency also operates the San Diego Coaster and Breeze Bus. Sprinter service is operated with Desiro-class diesel multiple units (DMU) manufactured by Siemens in Germany, where they are widely used by main-line regional railways. Twelve married pairs of Siemens VT642 Desiro DMUs were delivered to the Escondido Transit Center in August 2006. The vehicles were in acceptance testing in California during the early part of 2007. At Oceanside Transit Center, the Sprinter connects to three commuter rail lines (the Coaster, the Metrolink Orange County Line, and the Metrolink Inland Empire-Orange County Line) and Amtrak's Pacific Surfliner regional rail line.

History

The Sprinter is the first passenger train service of along the Escondido Branch since the Santa Fe Railroad discontinued passenger service in 1946. Originally built in 1888, the entire line had to be rebuilt to accommodate more traffic and be elevated because of the line runs along a river.[6] The Escondido and Vista stations are the only surviving passenger stations from the earlier era of service.

The funding for Sprinter originated with the TransNet Tax (Proposition C) measure passed by San Diego County voters in 1987 to relieve traffic congestion. A third of the tax was dedicated to mass transit.[7] [8] The $477 million project also was funded through a $152 million Full Funding Grant Agreement from the Federal Transit Administration.[6]

North County Transit District purchased the line in 1992 from the Santa Fe Railway. Construction started on the line in 2005.[6] and was scheduled for completion in December 2007. The Sprinter was scheduled to be previewed on December 28, 2007, with full runs beginning on January 13, 2008.[9]

However, the line was delayed until January 27 due to safety concerns.[10] Many doubted the reliability of the safety gates at level crossings. The reliability of the safety gates was associated with poor rail conditions and the Desiro-class diesel multiple units being lighter than the freight trains.

After months of use, rail conditions generally improve through grooving of the rail.[10] On January 25, it was announced that the Sprinter's opening was postponed again, this time to March 9.[11] The delays were caused by multiple issues. The California Public Utilities Commission did not certify the rail line until March 7, 2008, just 2 days before service began. It was also reported by North County Times that part of the problem involved the platforms at Escondido Avenue Station.

Due to its shared right-of-way with freight trains serving businesses in Escondido, the Sprinter platforms had to be set back from the tracks a sufficient distance to provide enough room for employees riding on the sides of freight cars. The passenger trains are not FRA compliant for operation in association with freight trains and therefore freight operations on the route are not permitted during passenger operations. For this reason some publications refer to this line as light rail but it does not conform with the usual understanding of that term.

While the DMUs aren't much narrower than freight cars, the space for employees hanging at the sides of cars considerably increases the free space required, and gangways were designed into the station that fold up after end of service to allow the BNSF trains plus employees at their sides to pass through. At the eastbound side of the Escondido Avenue platform, the tracks curve so sharply that a gap exists between the outside edges of the gangway and the side of the DMU. The California Public Utilities Commission has stated that such a gap is unsafe, and as a result, the Eastbound platform at Escondido Avenue was not used for six months after the opening of the SPRINTER. On Friday, September 12, 2008, the station was completed and on Monday, September 15, 2008, the station became operational.

Criticism

The Sprinter has encountered some dissatisfaction in northern San Diego County. For example, business owners in Oceanside have attributed flooding to Sprinter, since its construction raised railroad beds and narrowed creeks. The flooding occurred in November 2007 and January 2008.[12]

Some have also criticized the limited schedule, especially since the last westbound trains for Oceanside depart Cal State San Marcos and Palomar College shortly after 9:00 p.m. on weeknights.[13]

Ridership numbers first climbed after opening to great fanfare. According to a North County Times article, the Sprinter's ridership numbers reached just under 8,000 people per day, as of March 20.[14]

In the months following, the ridership decreased to the "mid 6,000s range" by May, 2008,[15] but has rebounded to near 8,000 per day by October, 2008.[1]

SANDAG's 2007 study predicted that the Sprinter would reduce road trips by 5,000 a day (a round trip by car would be two road trips). It also predicted over 11,000 riders (trips) per day by the end of the first year.[16]

Despite criticism over the budget and schedule Sprinter was the least expensive project per mile of 10 other rail projects built or planned in California in 2005.[17] American Public Works Association (APWA) awarded Sprinter the Transportation Project of the Year for projects valued over $75 million.[17]

Compass Card

In September 2008, SANDAG introduced a new contactless "Compass Card", made possible by Cubic Transportation Systems, Inc.. The "Compass Card" allows passengers from MTS and NCTD to store regional transit passes and cash value on a rewriteable RFID card. Customers can purchase passes and add cash value on the Internet or at any ticket vending machine. When a customer boards a bus they simply tap their Compass Card on the "Validator", usually located near or on the farebox. The LED display then lights up with lights resembling that of a stoplight, and the LCD display shows text regarding the passenger's fare account. When boarding rail vehicles the Validators are located on the platform, and the same process is performed to board the train.

Accidents

On March 11, 2008, just two days after the first passengers were carried, a westbound Sprinter train struck a man who was lying on the tracks under a State Route 78 bridge in San Marcos. It was not immediately clear if the man was aware of the approach of the train. However, the man, who was covered by a sleeping bag at the time he was struck, spoke of suicide while in the emergency room.[18]

Gallery

Sprinter at Oceanside, 2007-11-03.

Sprinter light rail vehicle, 2008

See also

- Transportation in San Diego County
- Press Article from San Diego Union Tribune [19]

External links

- Official site [20]
- The Sprinter in service (video) [21]

References

[1] Sisson, Paul (2008-10-30). "Sprinter ridership hovers around 8,000 per day" (http://www.nctimes.com/articles/2008/10/30/news/coastal/oceanside/zbd696e10915cd0f7882574f200623c1c.txt). *North County Times*. . Retrieved 2008-10-30.

[2] Burge, Michael (2008-03-10). "Passenger train sprints into service" (http://www.signonsandiego.com/news/northcounty/20080310-9999-1m10sprint.html). *San Diego Union-Tribune*. . Retrieved 2008-03-10.

[3] North County Transit District, The SPRINTER Commemorates One Year of Service (http://www.gonctd.com/about_news_09.htm) (press release), March 9, 2009: "Today marks the first full year of operations for the SPRINTER light rail train."

[4] Trains, Sprinter reaches the finish line, June 2008, p. 28: "North County Transit District's 22-mile Sprinter diesel light rail line"

[5] "North County Transit District-We Move People" (http://www.gonctd.com/sprinter_stations.htm). Gonctd.com. . Retrieved 2011-04-26.

[6] Angela Cotey (January 2008). "San Diego's North County Transit District launches SPRINTER light-rail service" (http://www.progressiverailroading.com/pr/article/Full-speed-ahead-for-San-Diegos-SPRINTER--14634). .

[7] "Transnet Tax for Congestion Relief" (http://thumper.tmisnet.com/~strads/transnet/transnet.html). Thumper.tmisnet.com. . Retrieved 2011-04-26.

[8] http://www.sandiego.gov/fm/annual/pdf/fy05/06v1glance.pdf

[9] "Sprinter (North County Transit District) - Wikipedia, the free encyclopedia" (http://www.gonctd.com/sprinter_schedules.htm). Gonctd.com. . Retrieved 2011-04-26.

[10] Burge, Michael (2008-01-18). "Sprinter could start on Jan. 27" (http://www.signonsandiego.com/news/northcounty/20080118-9999-1mc18nctd.html). *San Diego Union-Tribune*. . Retrieved 2008-03-10.

[11] Burge, Michael (2008-01-25). "Sprinter won't ride the rails until March 9" (http://www.signonsandiego.com/news/northcounty/20080125-1903-bn25sprint.html). *San Diego Union-Tribune*. . Retrieved 2008-03-10.

[12] Sherman, Lola (2008-01-09). "Businesses flooded along Sprinter rail line" (http://www.signonsandiego.com/news/northcounty/20080109-9999-1m9flood.html). *San Diego Union-Tribune*. . Retrieved 2008-03-10.

[13] Kovrig, Neill (2008-01-15). "Sprinter already a disappointment" (http://www.nctimes.com/articles/2008/01/16/opinion/commentary/21_30_191_15_08.txt). *North County Times*. . Retrieved 2008-03-10.

[14] Sisson, Paul (2008-03-20). "Sprinter ridership nudges toward 8,000 passengers" (http://www.nctimes.com/articles/2008/03/21/news/top_stories/21_27_403_20_08.txt). *North County Times*. . Retrieved 2008-06-09.

[15] UTU: News (http://www.utu.org/print_news.cfm?ArticleID=42481)

[16] "the rEgion: SANDAG's Electronic Newsletter" (http://www.sandag.org/enewsletter/archives/november2007/feature_7.html). Sandag.org. 2008-01-13. . Retrieved 2011-04-26.

[17] By CHRIS NICHOLS - cnichols@nctimes.com (2009-12-12). "REGION: Sprinter wins 'Project of Year' award" (http://www.nctimes.com/news/local/sdcounty/article_0755e088-1717-5314-a144-899490be488c.html). Nctimes.com. . Retrieved 2011-04-26.

[18] Burge, Michael (2008-03-12). "New Sprinter train hits man lying on tracks; legs severed" (http://www.signonsandiego.com/news/northcounty/20080312-9999-1m12sprint.html). San Diego Union-Tribune. . Retrieved 2008-04-09.

[19] Burge, Michael (2007-12-29). "News - North County - All aboard for Sprinter test run" (http://www.signonsandiego.com/news/northcounty/20071229-9999-1mi29sprint.html). SignOnSanDiego.com. . Retrieved 2011-04-26.

[20] http://www.gonctd.com/sprinter_intro.htm

[21] http://www.vimeo.com/1241441

Diesel_locomotive

A **diesel locomotive** is a type of railroad locomotive in which the prime mover is a diesel engine, a reciprocating engine operating on the Diesel cycle as invented by Dr. Rudolf Diesel. Several types of diesel locomotive have been developed, the principal distinction being in the means by which the prime mover's mechanical power is conveyed to the driving wheels (drivers).

Three styles of **diesel locomotive** body: cab unit, hood unit and box cab. These locomotives are operated by Pacific National in Australia.

Overview

Early internal combustion engine-powered locomotives used gasoline as their fuel. Soon after Dr. Rudolf Diesel patented his first compression ignition engine in 1892,[1] its application for railway propulsion was considered. Progress was slow, however, because of the poor power-to-weight ratio of the early engines, as well as the difficulty inherent in mechanically applying power to multiple driving wheels on swiveling trucks (bogies).

The InterCity 125, the current confirmed record holder as the fastest diesel-powered train at 148 mph (238 km/h); is made up of two power cars, one at each end of a fixed formation of carriages; capable of 125 mph (201 km/h) in regular service.

Steady improvements in the diesel engine's design (many developed by Sulzer Ltd. of Switzerland, with whom Dr. Diesel was associated for a time) gradually reduced its physical size and improved its power-to-weight ratio to a point where one could be mounted in a locomotive. Once the concept of Diesel-electric drive was accepted the pace of development quickened, and by 1925 a small number of diesel locomotives of 600 horsepower were in service in the United States. In 1930, Armstrong Whitworth of the United Kingdom delivered two 1,200 hp locomotives, using engines of Sulzer design, to Buenos Aires Great Southern Railway of Argentina.

By the mid 1950s, with economic recovery from the Second World War, series production of diesel locomotives had begun in many countries and the diesel locomotive was on its way to become the dominant type of locomotive in many countries and regions, offering greater flexibility and performance than the steam locomotive, as well as substantially lower operating and maintenance costs, other than where electric traction was in use due to policy decisions. Currently, almost all diesel locomotives are diesel-electric, although the diesel-hydraulic type was also widely used between the 1950s and 1970s.

Twin-section diesel locomotive 2M62M-1198 (rebuilt with CAT engines), near Kyviškės, Lithuania.

History

Adaptation of the diesel engine for rail use

The world's first oil-engined railway locomotive was built for the Royal Arsenal, Woolwich, England, in 1896, using an engine designed by Herbert Akroyd Stuart.[2] It was not, strictly, a diesel because it used a hot bulb engine (also known as a semi-diesel) but it was the precursor of the diesel.

A WDM-3A diesel locomotive of Indian Railways, used to haul both passenger and freight.

Following the expiration of Dr. Rudolph Diesel's patent in 1912, his engine design was successfully applied to marine propulsion and stationary applications. However, the massiveness and poor power-to-weight ratio of these early engines made them unsuitable for propelling land-based vehicles. Therefore, the engine's potential as a railroad prime mover was not initially recognized.[3] This changed as development reduced the size and weight of the engine.

The world's first diesel-powered locomotive was operated in the summer of 1912 on the Winterthur-Romanshorn Railroad in Switzerland, but was not a commercial success.[4] In 1906, Rudolf Diesel, Adolf Klose and the steam and Diesel engine manufacturer Gebrüder Sulzer founded Diesel-Sulzer-Klose GmbH for the manufacture of Diesel-powered locomotives. Sulzer had been manufacturing Diesel engines since 1898. Prussian State Railways ordered in 1909 a Diesel locomotive from the company, and after test runs between Winterthur and Romanshorn the Diesel-mechanical locomotive was delivered in Berlin in September 1912. During further test runs in 1913 several problems were found. After the first world war broke out in 1914, all further trials were stopped. The locomotive weight was 95 tonnes and the power was 883 kW with a maximum speed of 100 km/h.[5] Small numbers of prototype diesel locomotives were produced in a number of countries through the mid 1920s.

Early American developments

Adolphus Busch purchased the American manufacturing rights for the Diesel engine in 1898 but never applied this new form of power to transportation. Only limited success was achieved in the early twentieth century with direct-driven gasoline and Diesel powered railcars.[6]

General Electric (GE) entered the railcar market in the early twentieth century, as Thomas Edison possessed an outstanding patent on the electric locomotive, his design actually being a type of electrically propelled railcar.[7] GE built its first electric locomotive prototype in 1895. However, high electrification costs caused GE to turn its attention to Diesel power to provide electricity for electric railcars. Problems related to co-coordinating the Diesel engine and electric motor were immediately encountered, primarily due to limitations of the Ward Leonard electric elevator drive system that had been chosen.

A significant breakthrough occurred in 1914, when Hermann Lemp, a GE electrical engineer, developed and patented a reliable direct current electrical control system (subsequent improvements were also patented by Lemp).[8] Lemp's design used a single lever to control both engine and generator in a coordinated fashion, and was the prototype for all diesel-electric locomotive control systems.

In 1917–18, GE produced three experimental diesel-electric locomotives using Lemp's control design, the first known to be built in the United States.[9] Following this development, the 1923 Kaufman Act banned steam locomotives from New York City because of severe pollution problems. The response to this law was to electrify high-traffic rail lines. However, electrification was uneconomical to apply to lower-traffic areas.

The first regular use of diesel-electric locomotives was in switching (shunter) applications. General Electric produced several small switching locomotives in the 1930s (the famous "44-tonner" switcher was introduced in 1940) Westinghouse Electric and Baldwin collaborated to build switching locomotives starting in 1929. However, the Great Depression curtailed demand for Westinghouse's electrical equipment, and they stopped building locomotives internally, opting to supply electrical parts instead.[10]

First American series production locomotives

General Electric continued to be interested in developing a practical diesel railway locomotive, and approached Ingersoll-Rand in 1924. The resulting 300 horsepower locomotive was fitted with an electrical generator and traction motors supplied by GE, as well as a form of Lemp's control system, and was delivered in July 1925. This locomotive demonstrated that the diesel-electric power unit could provide many of the benefits of an electric locomotive without the railroad having to bear the sizeable expense of electrification.[11] The unit successfully demonstrated—in switching, road freight and passenger service—on a baker's dozen of railroads, and became the prototype for 33 600 horsepower AGEIR boxcab switching locomotives built by a consortium of GE, I-R and the American Locomotive Company for several New York City railroads.[12]

In June 1925, Baldwin Locomotive Works outshopped a prototype diesel-electric locomotive for "special uses" (such as for runs where water for steam locomotives was scarce) using electrical equipment from Westinghouse Electric Company.[13] Its twin-engine design was not successful, and the unit was scrapped after a short testing and demonstration period.[14] Industry sources were beginning to suggest "the outstanding advantages of this new form of motive power."[15] In 1929, the Canadian National Railways became the first North American railway to use diesels in mainline service with two units, 9000 and 9001, from Westinghouse.[16]

Diesel-electric railroad locomotion entered the American mainstream when the Burlington Railroad and Union Pacific used Diesel "streamliners" to haul passengers.[6] [17] Following the successful 1939 tour of General Motors' EMD's FT demonstrator freight locomotive set, the transition from steam to Diesel power began, the pace substantially quickening in the years following the close of World War II. Fairbanks-Morse developed a unique opposed-piston engine that was used in their locomotives, as well as in submarines.[18]

Early diesel-electric locomotives in the United States used direct current (DC) traction motors, but alternating current (AC) motors came into widespread use in the 1990s, starting with the Electromotive SD70MAC in 1993 and followed by the General Electric's AC4400CW in 1994 and AC6000CW in 1995.[19]

Early diesel locomotives and railcars outside the United States

In 1935, Krauss-Maffei, MAN and Voith built the first diesel-hydraulic locomotive, the V140 in Germany. Because the German railways (DR) was very pleased with the performance of the locomotive, diesel-hydraulics became the mainstream in diesel locomotives in Germany. Series production of diesel locomotives in Germany began after the Second World War.

Diesel-powered or "oil-engined" railcars, generally diesel-mechanical, were developed by various European manufacturers in the 1930s, e.g. by William Beardmore and Company for the Canadian National Railways (the Beardmore Tornado engine was subsequently used in the R101 airship), the Flying Hamburger two coach diesel railcar which entered service in Germany in 1933, and the Vulcan railcars of 1940 for New Zealand.

In 1945 a batch of 30 Baldwin diesel-electric locomotives, Baldwin 0-6-6-0 1000, was delivered from the United States to the railways of the Soviet Union, although a first Russian diesel-electric locomotive (designed by a team led by Yuri Lomonosov) is reported to have entered service in 1924.

Fiat claims a first Italian diesel-electric locomotive built in 1922, but little detail is available. A Fiat-TIBB diesel-locomotive "A", of 440CV, is reported to have entered service on the Ferrovie Calabro Lucane in southern Italy in 1926, following trials in 1924-5. http://www.ferrovie.it/forum/viewtopic.php?f=22&t=13653 Series production of diesel locomotives in Italy began in the mid 1950s.

In 1948, the London Midland & Scottish Railway introduced the first of a pair of 1,600 hp Co-Co diesel-electric locomotives (later British Rail Class D16/1) for regular use in the United Kingdom, although British manufacturers such as Armstrong Whitworth had been exporting diesel locomotives since 1930. Fleet deliveries to British Railways, of other designs such as Class 20 and Class 31, began in 1957.

Series production of China's first diesel locomotive class, the DFH 1, began in 1964 following construction of a prototype in 1959.

Diesel's advantages over steam

Diesel engines slowly eclipsed those powered by steam as the manufacturing and operational efficiencies of the former made them cheaper to own and operate. While initial costs of diesel engines were high, steam locomotives were custom-made for specific railway routes and lines and, as such, economies of scale were difficult to achieve.[20] Though more complex to produce with exacting manufacturing tolerances ($^1/_{10000}$-inch (0.0025 mm) for diesel, compared with $^1/_{100}$-inch (0.25 mm) for steam), diesel locomotive parts were more conducive to mass production. While the steam engine manufacturer Baldwin offered almost five hundred steam models in its heyday, EMD offered fewer than ten diesel varieties.[21]

Diesel locomotives offer significant operating advantages over steam locomotives.[22] They can safely be operated by one person, making them ideal for switching/shunting duties in yards (although for safety reasons many main-line diesel locomotives continue to have 2-man crews) and the operating environment is much more attractive, being much quieter, fully weatherproof and without the dirt and heat that is an inevitable part of operating a steam locomotive.

Diesel engines can be started and stopped almost instantly, meaning that a diesel locomotive has the potential to incur no costs when not being used. However, it is still the practice of large North American railroads to use straight water as a coolant in diesel engines instead of coolants that incorporate anti-freezing properties; this results in diesel locomotives being left idling when parked in cold climates instead of being completely shut down. Still, a diesel engine can be left idling unattended for hours or even days, especially since practically every diesel engine used in locomotives has systems that automatically shut the engine down if a problem such as a loss of oil pressure or coolant loss occur.

Steam locomotives, by comparison, require intensive maintenance, lubrication and cleaning before, during and after use. Preparing and firing a steam locomotive for use from cold can take many hours, although it may be kept in readiness between uses with a small fire to maintain a slight heat in the boiler, but this requires regular stoking and frequent attention to maintain the level of water in the boiler. This may be necessary to prevent the water in the boiler freezing in cold climates, so long as the water supply itself is not frozen.

Moreover, maintenance and operational costs of steam locomotives were much higher than diesel counterparts even though it would take diesel locomotives almost 50 years to reach the same power output that steam locomotives could achieve at their technological height. Annual maintenance costs for steam locomotives accounted for 25% of the initial purchase price. Spare parts were cast from wooden masters for specific locomotives. The sheer number of unique steam locomotives meant that there was no feasible way for spare-part inventories to be maintained.[23]

Steam engines also required large quantities of coal and water, which were expensive variable operating costs.[24] Further, the thermal efficiency of steam was considerably less than that of diesel engines. Diesel's theoretical studies demonstrated potential thermal efficiencies for a compression ignition engine of 36% (compared with 6-10% for steam), and a 1897 one-cylinder prototype operated at a remarkable 26% efficiency.[25]

The Trans-Australian Railway built 1912 to 1917 by Commonwealth Railways (CR) passes through 2000 km of waterless (or salt watered) desert terrain unsuitable for steam locomotives. The original engineer Henry Deane envisaged diesel operation to overcome such problems.[26] Some have suggested that the CR worked with the South Australian Railways to trial diesel traction.[27] However, the technology was not developed enough to be reliable.

By the mid 1960s, diesel locomotives had effectively replaced steam engines where electric traction was not in use.[24]

Transmission types

Unlike steam engines, internal combustion engines require a transmission to power the wheels. The engine must be allowed to continue to run when the locomotive is stopped.

Diesel-mechanical

A diesel-mechanical locomotive uses a mechanical transmission in a fashion similar to that employed in most road vehicles. This type of transmission is generally limited to low-powered, low speed shunting (switching) locomotives, lightweight multiple units and self-propelled railcars.

A British Rail Class 03 Diesel-mechanical shunter (switcher) with a jackshaft under the cab.

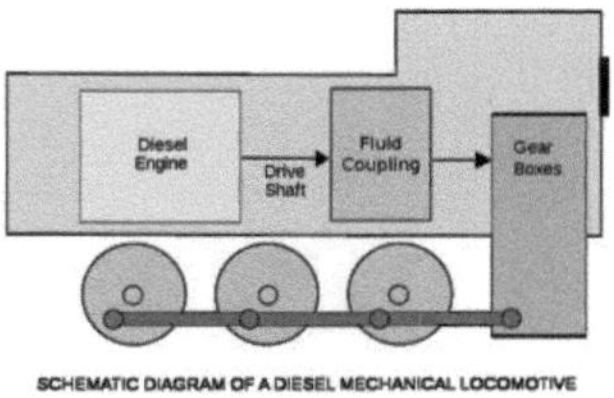

Schematic illustration of a diesel mechanical locomotive

The mechanical transmissions used for railroad propulsion are generally more complex and much more robust than road versions. There is usually a fluid coupling interposed between the engine and gearbox, and the gearbox is often of the epicyclic (planetary) type to permit shifting while under load. Various systems have been devised to minimise the break in transmission during gear changing, e.g. the S.S.S. (synchro-self-shifting) gearbox used by Hudswell Clarke.

Diesel-mechanical propulsion is limited by the difficulty of building a reasonably sized transmission capable of coping with the power and torque required to move a heavy train. A number of attempts to use Diesel-mechanical propulsion in high power applications have been made (e.g. the 1,500 kW (2000 horsepower) British Rail 10100 locomotive), although none have proved successful in the end.

Diesel-electric

For locomotives powered by both external electricity and diesel fuel, see electro-diesel below. For locomotives powered by a combination of diesel or fuel cells and batteries or ultracapacitors, see hybrid locomotive.

In a **Diesel-electric** locomotive, the Diesel engine drives an electrical generator whose output provides power to the traction motors. There is no mechanical connection between the engine and the wheels. The important components of Diesel-electric propulsion are the Diesel engine (also known as the prime mover), the main generator, traction motors and a control system consisting of the engine governor as well as electrical and/or electronic components used to control or modify the electrical supply to the traction motors, including switchgear, rectifiers and other components. In the most elementary case, the generator may be directly connected to the motors with only very simple switchgear.

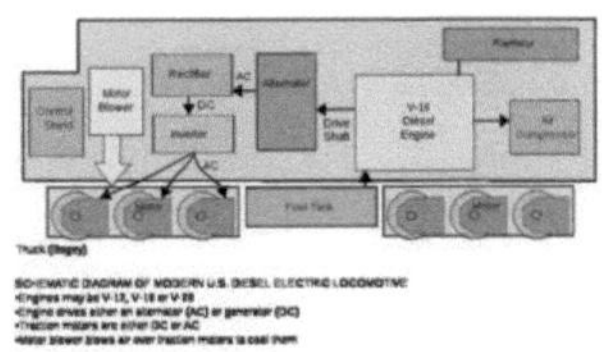

Schematic diagram of Diesel electric locomotive

the EMD F40PH (left) and MPI MPXpress-series MP36PH-3S (right) locomotives coupled together by Metra use Diesel-electric transmission.

Originally, the traction motors and generator were DC machines. Following the development of high-capacity silicon rectifiers in the 1960s, the DC generator was replaced by an alternator using a diode bridge to convert its output to DC. This advance greatly improved locomotive reliability and decreased generator maintenance costs by elimination of the commutator and brushes in the generator. Elimination of the brushes and commutator, in turn, disposed of the possibility of a particularly destructive type of event referred to as a flashover, which could result in immediate generator failure and, in some cases, start an engine room fire.

More recently, the development of high-power variable-frequency/variable-voltage (VVVF) drives, or "traction inverters," has allowed the use of polyphase AC traction motors, thus also eliminating the motor commutator and brushes. The result is a more efficient and reliable drive that requires relatively little maintenance and is better able to cope with overload conditions that often destroyed the older types of motors.

Soviet 2TE10U locomotive

Engineer's controls in a Diesel-electric locomotive cab. The lever near bottom-centre is the throttle and the lever visible at bottom left is the automatic brake valve control.

Diesel-electric control

MLW model S-3 produced in 1957 for the CPR adhering to designs by ALCO.

A Diesel-electric locomotive's power output is independent of road speed, as long as the unit's generator current and voltage limits are not exceeded. Therefore, the unit's ability to develop tractive effort (also referred to as drawbar pull or tractive force, which is what actually propels the train) will tend to inversely vary with speed within these limits. (See power curve below). Maintaining acceptable operating parameters was one of the principal design considerations that had to be solved in early Diesel-electric locomotive development and, ultimately, led to the complex control systems in place on modern units.

Throttle operation

The prime mover's power output is primarily determined by its rotational speed (RPM) and fuel rate, which are regulated by a governor or similar mechanism. The governor is designed to react to both the throttle setting, as determined by the engine driver and the speed at which the prime mover is running.[28]

An EMD 12-567B 12 cylinder Diesel engine, stored pending rebuild, and missing some components.

Locomotive power output, and thus speed, is typically controlled by the engine driver using a stepped or "notched" throttle that produces binary-like electrical signals corresponding to throttle position. This basic design lends itself well to multiple unit (MU) operation by producing discrete conditions that assure that all units in a consist respond in the same way to throttle position. Binary encoding also helps to minimize the number of trainlines (electrical connections) that are required to pass signals from unit to unit. For example, only four trainlines are required to encode all throttle positions.

North American locomotives, such as those built by EMD or General Electric, have nine throttle positions, one idle and eight power (as well as an emergency stop position that shuts down the prime mover). Many UK-built locomotives have a ten-position throttle. The power positions are often referred to by locomotive crews as "run 3" or "notch 7", depending upon the throttle setting.

In older locomotives, the throttle mechanism was ratcheted so that it was not possible to advance more than one power position at a time. The engine driver could not, for example, pull the throttle from notch 2 to notch 4 without stopping at notch 3. This feature was intended to prevent rough train handling due to abrupt power increases caused by rapid throttle motion ("throttle stripping," an operating rules violation on many railroads). Modern locomotives no longer have this restriction, as their control systems are able to smoothly modulate power and avoid sudden changes in train loading regardless of how the engine driver operates the controls.

When the throttle is in the idle position, the prime mover will be receiving minimal fuel, causing it to idle at low RPM. In addition, the traction motors will not be connected to the main generator and the generator's field windings will not be excited (energized) — the generator will not produce electricity with no excitation. Therefore, the locomotive will be in "neutral". Conceptually, this is the same as placing an automobile's transmission into neutral while the engine is running.

To set the locomotive in motion, the reverser control handle is placed into the correct position (forward or reverse), the brake is released and the throttle is moved to the run 1 position (the first power notch). An experienced engine driver can accomplish these steps in a coordinated fashion that will result in a nearly imperceptible start. The positioning of the reverser and movement of the throttle together is conceptually like shifting an automobile's automatic transmission into gear while the engine is idling

Placing the throttle into the first power position will cause the traction motors to be connected to the main generator and the latter's field coils to be excited. With excitation applied, the main generator will deliver electricity to the traction motors, resulting in motion. If the locomotive is running "light" (that is, not coupled to a train) and is not on an ascending grade, it will easily accelerate. On the other hand, if a long train is being started, the locomotive may stall as soon as some of the slack has been taken up, as the drag imposed by the train will exceed the tractive force being developed. An experienced engine driver will be able to recognize an incipient stall and will gradually advance the throttle as required to maintain the pace of acceleration.

As the throttle is moved to higher power notches, the fuel rate to the prime mover will increase, resulting in a corresponding increase in RPM and horsepower output. At the same time, main generator field excitation will be proportionally increased to absorb the higher power. This will translate into increased electrical output to the traction motors, with a corresponding increase in tractive force. Eventually, depending on the requirements of the train's schedule, the engine driver will have moved the throttle to the position of maximum power and will maintain it there until the train has accelerated to the desired speed.

As will be seen in the following discussion, the propulsion system is designed to produce maximum traction motor torque at start-up, which explains why modern locomotives are capable of starting trains weighing in excess of 15,000 tons, even on ascending grades. Current technology allows a locomotive to develop as much as 30 percent of its loaded driver weight in tractive force, amounting to some 120000 pounds (54000 kg) of drawbar pull for a large, six-axle freight (goods) unit. In fact, a consist of such units can produce more than enough drawbar pull at start-up to damage or derail cars (if on a curve) or break couplers (the latter being referred to in North American railroad slang as "jerking a lung"). Therefore, it is incumbent upon the engine driver to carefully monitor the amount of power being applied at start-up to avoid damage. In particular, "jerking a lung" could be a calamitous matter if it were to occur on an ascending grade.

Propulsion system operation

As previously explained, the locomotive's control system is designed so that the main generator electrical power output is matched to any given engine speed. Given the innate characteristics of traction motors, as well as the way in which the motors are connected to the main generator, the generator will produce high current and low voltage at low locomotive speeds, gradually changing to low current and high voltage as the locomotive accelerates. Therefore, the net power produced by the locomotive will remain constant for any given throttle setting (*see power curve graph for notch 8*).

In older designs, the prime mover's governor and a companion device, the load regulator, play a central role in the control system. The governor has two external inputs: requested engine speed, determined by the engine driver's throttle setting, and actual engine speed (feedback). The governor has two external control outputs: fuel injector setting, which determines the engine fuel rate, and load regulator position, which affects main generator excitation. The governor also incorporates a separate overspeed protective mechanism that will immediately cut off the fuel supply to the injectors and sound an alarm in the cab in the event the prime mover exceeds a defined RPM. Not all of these inputs and outputs are necessarily electrical.

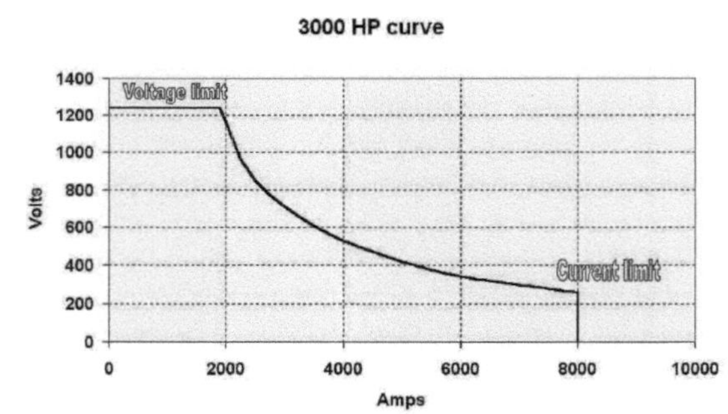

Typical main generator constant power curve at "notch 8".

The load regulator is essentially a large potentiometer that controls the main generator power output by varying its field excitation and hence the degree of loading applied to the engine. The load regulator's job is relatively complex, because although the prime mover's power output is proportional to RPM and fuel rate, the main generator's output is not (which characteristic was not correctly handled by the Ward Leonard elevator drive system that was initially tried in early locomotives).

As the load on the engine changes, its rotational speed will also change. This is detected by the governor through a change in the engine speed feedback signal. The net effect is to adjust both the fuel rate and the load regulator position so that engine RPM and torque (and thus power output) will remain constant for any given throttle setting, regardless of actual road speed.

In newer designs controlled by a “traction computer,” each engine speed step is allotted an appropriate power output, or “kW reference”, in software. The computer compares this value with actual main generator power output, or “kW feedback”, calculated from traction motor current and main generator voltage feedback values. The computer adjusts the feedback value to match the reference value by controlling the excitation of the main generator, as described above. The governor still has control of engine speed, but the load regulator no longer plays a central role in this type of control system. However, the load regulator is retained as a “back-up” in case of engine overload. Modern locomotives fitted with electronic fuel injection (EFI) may have no mechanical governor, however a “virtual” load regulator and governor are retained with computer modules.

Traction motor performance is controlled either by varying the DC voltage output of the main generator, for DC motors, or by varying the frequency and voltage output of the VVVF for AC motors. With DC motors, various connection combinations are utilized to adapt the drive to varying operating conditions.

At standstill, main generator output is initially low voltage/high current, often in excess of 1000 amperes per motor at full power. When the locomotive is at or near standstill, current flow will be limited only by the DC resistance of the motor windings and interconnecting circuitry, as well as the capacity of the main generator itself. Torque in a series-wound motor is approximately proportional to the square of the current. Hence, the traction motors will produce their highest torque, causing the locomotive to develop maximum tractive effort, enabling it to overcome the inertia of the train. This effect is analogous to what happens in an automobile automatic transmission at start-up, where it is in first gear and thus producing maximum torque multiplication.

As the locomotive accelerates, the now-rotating motor armatures will start to generate a counter-electromotive force (back EMF, meaning the motors are also trying to act as generators), which will oppose the output of the main generator and cause traction motor current to decrease. Main generator voltage will correspondingly increase in an attempt to maintain motor power, but will eventually reach a plateau. At this point, the locomotive will essentially cease to accelerate, unless on a downgrade. Since this plateau will usually be reached at a speed substantially less than the maximum that may be desired, something must be done to change the drive characteristics to allow continued acceleration. This change is referred to as "transition," a process that is analogous to shifting gears in an automobile.

Transition methods include:

- Series / Parallel or "motor transition".
  - Initially, pairs of motors are connected in series across the main generator. At higher speed, motors are reconnected in parallel across the main generator.
- "Field shunting", "field diverting", or "weak fielding".
  - Resistance is connected in parallel with the motor field. This has the effect of increasing the armature current, producing a corresponding increase in motor torque and speed.

Both methods may also be combined, to increase the operating speed range.

- Generator transition
  - Reconnecting the two separate internal main generator stator windings from parallel to series to increase the output voltage.

In older locomotives, it was necessary for the engine driver to manually execute transition by use of a separate control. As an aid to performing transition at the right time, the load meter (an indicator that informs the engine driver on how much current is being drawn by the traction motors) was calibrated to indicate at which points forward or backward transition should take place. Automatic transition was subsequently developed to produce better operating efficiency, and to protect the main generator and traction motors from overloading from improper transition.

Dynamic braking

A common option on Diesel-electric locomotives is dynamic (rheostatic) braking.

Dynamic braking takes advantage of the fact that the traction motor armatures are always rotating when the locomotive is in motion and that a motor can be made to act as a generator by separately exciting the field winding. When dynamic braking is utilized, the traction control circuits are configured as follows:

- The field winding of each traction motor is connected across the main generator.
- The armature of each traction motor is connected across a forced-air cooled resistance grid (the dynamic braking grid) in the roof of the locomotive's hood.
- The prime mover RPM is increased and the main generator field is excited, causing a corresponding excitation of the traction motor fields.

The aggregate effect of the above is to cause each traction motor to generate electric power and dissipate it as heat in the dynamic braking grid. A fan connected across the grid provides forced-air cooling. Consequently, the fan is powered by the output of the traction motors and will tend to run faster and produce more airflow as more energy is applied to the grid.

Ultimately, the source of the energy dissipated in the dynamic braking grid is the motion of the locomotive as imparted to the traction motor armatures. Therefore, the traction motors impose drag and the locomotive acts as a brake. As speed decreases, the braking effect decays and usually becomes ineffective below approximately 16 km/h (10 mph), depending on the gear ratio between the traction motors and axles.

Dynamic braking is particularly beneficial when operating in mountainous regions, where there is always the danger of a runaway due to overheated friction brakes during descent (see also comments in the air brake article regarding loss of braking due to improper train handling). In such cases, dynamic brakes are usually applied in conjunction with the air brakes, the combined effect being referred to as blended braking. The use of blended braking can also assist in keeping the slack in a long train stretched as it crests a grade, helping to prevent a "run-in", an abrupt bunching of train slack that can cause a derailment. Blended braking is also commonly used with commuter trains to reduce wear and tear on the mechanical brakes that is a natural result of the numerous stops such trains typically make during a run.

Electro-diesel

Metro-North's GE Genesis P32AC-DM electro-diesel locomotive.

These special locomotives can operate as an electric locomotive or as a Diesel locomotive. The Long Island Rail Road and Metro-North Railroad operate dual-mode diesel-electric/third-rail locomotives between non-electrified territory and New York City because of a local law banning diesel-powered locomotives in Manhattan tunnels. For the same reason, Amtrak operates a fleet of dual-mode locomotives in the New York area. British Rail operated dual diesel-electric/electric locomotives designed to run primarily as electric locomotives with reduced power available when running on diesel power. This allowed railway yards to remain un-electrified, as the third rail power system is extremely hazardous in a yard area.

Diesel-hydraulic

Diesel-hydraulic locomotives use a torque converter or hydraulic drive system to convey the power from the diesel engine to the wheels.

Hydrokinetic transmission

Further information: torque converter

Hydrokinetic transmission (also called hydrodynamic transmission) uses a torque converter. A torque converter consists of three main parts, two of which rotate, and one that has a lock preventing backwards rotation and adding output torque by redirecting the oil flow at low output RPM. All three main parts are sealed in an oil-filled housing. To match engine speed to load speed over the entire speed range of a locomotive some additional method is required to give sufficient range. One method is to follow the torque converter with a mechanical gearbox which switches ratios automatically, similar to an automatic transmission on a car. Another method is to provide several torque converters each with a range of variability covering part of the total required; all the torque converters are mechanically connected all the time, and the appropriate one for the speed range required is selected by filling it with oil and draining the others. The filling and draining is carried out with the transmission under load, and results in very smooth range changes with no break in the transmitted power.

DB class V 200 diesel-hydraulic

Multiple units

Diesel-hydraulic multiple units, a less arduous duty, often use a simplification of this system, with a torque converter for the lower speed ranges and a fluid coupling for the high speed range. A fluid coupling is similar to a torque converter but it lacks the stator. The output torque is equal to the input torque regardless of the ratio of input to output speed; loading the output shaft results not in torque multiplication and constant power throughput but in reduction of the input speed with consequent lower power throughput. (In car terms, the fluid coupling provides top gear and the torque converter provides all the lower gears.) The result is that the power available at the rail is reduced when operating in the lower speed part of the fluid coupling range, but the less arduous duty of a passenger multiple unit compared to a locomotive makes this an acceptable trade-off for reduced mechanical complexity.

Locomotives

Diesel-hydraulic locomotives are less efficient than diesel-electrics. The first-generation BR diesel hydraulics were significantly less efficient (c. 65%) than diesel electrics (c. 80%)—but initial versions were found in many countries to be mechanically more complicated and more likely to break down. Hydraulic transmission for locomotives was developed in Germany. The bad reputation of diesel-hydraulic principle was caused by the poor durability and reliability of the Maybach Mekydro hydraulic transmission. The Mekydro consisted of a hydraulic torque converter followed by a four speed automatic mechanical gearbox. Voith developed a different solution using several torque converters, and it has proven to be extremely durable and very well-suited for the purpose.

British Rail diesel-hydraulic locomotives: Class 52 "Western", Class 42 "Warship" and Class 35 "Hymek"

Reliability

In Germany and Finland, diesel-hydraulic systems have achieved extremely high reliability in operation. Persistent argument continues over the relative reliability of hydraulic systems, with continuing questions over whether data were manipulated politically to favour local suppliers over German ones. In the US and Canada, they are now greatly outnumbered by diesel-electric locomotives, while they remain dominant in some European countries.

Examples

The most famous diesel-hydraulic locomotive is the German V 200, which were built from 1953 in a total number of 136. The only diesel-electric locomotives of the *Deutsche Bundesbahn* were BR 288 (V 188), of which the Deutsche Reichsbahn built 12 in 1939.

A VR Class Dv12 diesel-hydraulic locomotive

In the UK, the Western Region of British Rail bought a number of Diesel-hydraulic locomotives, ranging from small light-duty freight locomotives to high-powered mainline passenger locomotives, but these were withdrawn early because they were non-standard, and also in some cases suffering from reliability problems (see above), being replaced by diesel-electrics. A number were rescued for preservation though, and some are capable of running on the mainline.

A GMD GMDH-1 diesel-hydraulic locomotive

In Finland, over 200 Finnish-built VR class Dv12 and Dr14 diesel-hydraulics with Voith transmissions have been continuously used since the early 1960s. All units of Dr14 class and most units of Dv12 class are still in service. VR has abandoned some weak-conditioned units of 2700 series Dv12s.[29]

In the 1950s, General Motors Diesel of Canada built 4 GMD GMDH-1 experimental locomotives.

In 1959 the South African Railways (SAR) placed seven Henschel built diesel-hydraulic locomotives in service. Designated as South African Class 61-000, all but one were later sold to Rhodesia Railways, where they became RR Class DH1.

In the 1960s, the Southern Pacific (SP) bought 18 diesel-hydraulic locomotives KM ML-4000 from the Kraus-Maffei company. The Denver & Rio Grande Western also bought three, all of which it later sold to SP.[30] Only the outer shell of one of these (converted into a camera car by SP in the 1970s) exists today, the other 20 having been scrapped when SP terminated the experiment in 1970.

Hydrostatic transmission

Hydrostatic transmission is limited to small shunting locomotives and rail maintenance equipment. The diesel engine drives a hydraulic pump which supplies oil under pressure to hydraulic motors which drive the axles.

Diesel-steam

Steam-diesel hybrid locomotives can use steam generated from a boiler or diesel to power a piston engine. The *Cristiani Compressed Steam System* used a diesel engine to power a compressor to drive and recirculate steam produced by a boiler; effectively using steam as the power transmission medium, with the diesel engine being the prime mover[31]

Diesel-pneumatic

The diesel-pneumatic locomotive was of interest in the 1930s because it offered the possibility of converting existing steam locomotives to diesel operation. The frame and cylinders of the steam locomotive would be retained and the boiler would be replaced by a diesel engine driving an air compressor. The problem was low thermal efficiency because of the large amount of energy wasted as heat in the air compressor. Attempts were made to compensate for this by using the diesel exhaust to re-heat the compressed air but these had limited success. A German proposal of 1929 did result in a prototype[32] but a similar British proposal of 1932, to use an LNER Class R1 locomotive, never got beyond the design stage.

Multiple-unit operation

Diesel-electric locomotive built by EMD for service in the UK and continental Europe.

Most Diesel locomotives are capable of multiple unit operation (MU) as a means of increasing horsepower and tractive effort when hauling heavy trains. All North American locomotives, including export models, use a standardised AAR electrical control system interconnected by a 27-pin jumper cable between the units. For UK-built locomotives, a number of incompatible control systems are used, but the most common is the Blue Star system, which is electro-pneumatic and fitted to most early diesel classes. A small number of types, typically higher-powered locomotives intended for passenger only work, do not have multiple control systems. In all cases, the electrical control connections made common to all units in a consist are referred to as trainlines. The result is that all locomotives in a consist behave as one in response to the engine driver's control movements.

The ability to MU Diesel-electric locomotives was first introduced in the EMD FT four-unit demonstrator that toured the USA in 1939. At the time, American railroad work rules required that each operating locomotive in a train had to have on board a full crew. EMD circumvented that requirement by joining the individual units of the demonstrator with drawbars instead of conventional knuckle couplers and declaring the combination to be a single locomotive. Electrical interconnections were made so one engine driver could operate the entire consist from the head-end unit. Later on, work rules were amended and the semi-permanent joining of units with drawbars was eliminated in favour of couplers, as servicing had proved to be somewhat cumbersome owing to the total length of the consist (about 200 feet or nearly 61 meters).

In mountainous regions, it is common to interpose helper locomotives in the middle of the train, both to provide the extra power needed to ascend a grade and to limit the amount of stress applied to the draft gear of the car coupled to the head-end power. The helper units in such distributed power configurations are controlled from the lead unit's cab through coded radio signals. Although this is technically not MUing, the behaviour is the same as with physically interconnected units.

Cab arrangements

Cab arrangements vary by builder and operator. Practice in the U.S. has traditionally been for a cab at one end of the locomotive with limited visibility if the locomotive is not operated cab forward. This is not usually a problem as U.S. locomotives are usually operated in pairs, or threes, and arranged so that a cab is at each end of each set. European practice is usually for a cab at each end of the locomotive as trains are usually light enough to operate with one locomotive. Early U.S. practice was to add power units without cabs (booster or B-units) and the arrangement was often A-B, A-B-A, or A-B-B-A where A was a unit with a cab. Center cabs were sometimes used for switch locomotives.

Cow-calf

In North American railroading, a cow-calf set is a pair of switcher-type locomotives: one (the cow) equipped with a driving cab, the other (the calf) without a cab, and controlled from the cow through cables. Cow-calf sets are used in heavy switching and hump yard service. Some are radio controlled without an operating engineer present in the cab. This arrangement is also known as master-slave.

Flameproof diesel locomotive

A standard diesel locomotive presents a very low fire risk but "flameproofing" can reduce the risk even further. This involves fitting a water-filled box to the exhaust pipe to quench any red-hot carbon particles that may be emitted. Other precautions may include a fully insulated electrical system (neither side earthed to the frame) and all electric wiring enclosed in conduit.

The flameproof diesel locomotive has replaced the fireless steam locomotive in areas of high fire risk such as oil refineries and ammunition dumps. Preserved examples of flameproof diesel locomotives include:

- *Francis Baily of Thatcham* (ex-RAF Welford) at Southall Railway Centre
- *Naworth* (ex-National Coal Board) at the South Tynedale Railway[33]

Latest development of the "Flameproof Diesel Vehicle Applied New Exhaust Gas Dry Type Treatment System " does not need the water supply.[34]

Lights

The lights fitted to diesel locomotives vary from country to country. North American locomotives have a powerful headlight and a pair of ditch lights. The latter are fitted low down at the front and are designed to make the locomotive easily visible as it approaches a grade crossing. Older locomotives may be fitted with a Gyralite or Mars Light instead of the ditch lights.

A Canadian National Railway train showing the placement of the headlight and ditch lights on the locomotive.

Environmental impact

Although diesel locomotives pollute considerably less than steam locomotives, they still emit diesel exhaust that contributes to health effects (especially cardiovascular) and global warming.[35] In 2008, the United States Environmental Protection Agency (EPA) mandated regulations requiring all new or refurbished diesel locomotives to meet pollution standards that slash the amount of allowable soot by 90% and require an 80% reduction in nitrogen oxide emissions. For years, it was thought by government scientists who measure air pollution that diesel locomotive engines were relatively clean and emitted far less health-threatening emissions than that of diesel trucks or other vehicles; however, the scientists discovered that because they used faulty estimates of the amount of fuel consumed by diesel locomotives, they grossly understated the amount of pollution generated annually. After revising their calculations, they concluded that the annual emissions of nitrogen oxide, a major ingredient in smog, and soot would be by 2030 nearly twice what they originally assumed.[36] [37]

This would mean that diesel locomotives would be releasing more than 800,000 tons of nitrogen oxide and 25,000 tons of soot every year within a quarter of a century, in contrast to the EPA's previous projections of 480,000 tons of nitrogen dioxide and 12,000 tons of soot. Since this was discovered, to reduce the effects of the diesel locomotive on humans (who are breathing the noxious emissions) and on plants and animals, it is considered practical to install traps in the diesel engines to reduce pollution levels[38] and other forms (e.g., use of biodiesel).

Diesel locomotive pollution has been of particular concern in the city of Chicago. The *Chicago Tribune* reported levels of diesel soot inside locomotives leaving Chicago at levels hundreds of times above what is normally found on streets outside.[39] Residents of several neighborhoods are most likely exposed to diesel emissions at levels several times higher than the national average for urban areas.[40]

See also

- Diesel multiple unit
- Diesel-electric transmission
- Diesel engine
- Electric locomotive
- Electro-diesel locomotive
- Electrification
- Hybrid locomotive
- Hybrid electric vehicle

References

[1] Diesel, Rudolf. U.S. Patent No. 608,845, filed July 15, 1895, and issued August 9, 1898 *Accessed via Google Patent Search at: US Patent #608,845* (http://www.google.com/patents?vid=USPAT608845&id=vQVgAAAAEBAJ&dq=608845) *on February 8, 2007.*

[2] Doherty, J M 1962: *Diesel Locomotive Practice*. Odhams Press

[3] Churella, Albert J. (1998). *From Steam to Diesel: Managerial Customs and Organizational Capabilities in the Twentieth-Century American Locomotive Industry*. Princeton, New Jersey: Princeton University Press. p. 15. ISBN 0-691-02776-5.

[4] Churella, page 12

[5] Glatte, Wolfgang (1993). *Deutsches Lok-Archiv: Diesellokomotiven 4. Auflage*. Berlin: Transpress. ISBN 3-344-70767-1.

[6] Stover, John F. (1997). *American Railroads*. Chicago, Illinois: The University of Chicago Press. p. 212. ISBN 0-226-77658-I.

[7] Edison, Thomas A. U.S. Patent No. 493,425, filed January 19, 1891, and issued March 14, 1891 *Accessed via the Edison Papers at: US Patent #493,425* (http://edison.rutgers.edu/patents/00493425.PDF) *on February 8, 2007.*

[8] Lemp, Hermann. U.S. Patent No. 1,154,785, filed April 8, 1914, and issued September 28, 1915. *Accessed via Google Patent Search at: US Patent #1,154,785* (http://www.google.com/patents?vid=USPAT1154785&id=UhpBAAAAEBAJ&dq=1154785) *on February 8, 2007.*

[9] Pinkepank, Jerry A. (1973). *The Second Diesel Spotter's Guide*. Milwaukee WI: Kalmbach Books. pp. 139–141. ISBN 0-89024-026-4.

[10] Churella, pp. 28-30.

[11] Churella, pp. 25-27.

[12] Pinkepank, p. 209–211.

[13] "Railroads To Try Diesel Locomotive", *Special to the New York Times*: 1, February 18, 1925

[14] Pinkepank, p. 283.

[15] Churella, p. 27.

[16] Pinkepank, p. 409.

[17] "Diesel Streamliners Now Link Coast-to-Coast" (http://books.google.com/books?id=79oDAAAAMBAJ&pg=PA165&dq=Popular+Science+1933+plane+"Popular+Mechanics"&hl=en&ei=RasMTuyGFYifsQLC3sGzCg&sa=X&oi=book_result&ct=result&resnum=2&ved=0CC0Q6AEwATgK#v=onepage&q&f=true) *Popular Mechanics*, August 1937

[18] Wendel, C.H. (1987). *Power in the Past, Vol. 2; A History of Fairbanks-Morse and Co.*, Stemgas Publishing Company, 1982.

[19] Solomon, Brian, *Locomotive*, 2001, pp 120, 130

[20] Churella, 10

[21] Churella, 19

[22] (http://www.sdrm.org/faqs/hostling.html), Phil Jern "How to Boot a Steam Locomotive" (1990) San Diego Railroad Museum.

[23] Churella, 12-17

[24] Stover, 213

[25] Churella, 14

[26] Burke, A 1991., *Rails through the Wilderness*; New South Wales University Press

[27] Holden, R 2006 No. 259 : the curious story of a forgotten locomotive, Railmac Publications

[28] Control theory

[29] Suruliputus saatteli veturit viimeiselle matkalle (http://www.iltasanomat.fi/uutiset/kotimaa/uutinen.asp?id=1740927) **(Finnish)**

[30] Marre, Louis A. (1995). *Diesel Locomotives: The First Fifty Years*. Waukesha, Wis., USA: Kalmbach. pp. 384–385. ISBN 0890242585.

[31] The Paragon-Cristiani Compressed Steam System (http://www.aqpl43.dsl.pipex.com/MUSEUM/LOCOLOCO/compsteam/compsteam.htm) *dslef.dsl.pipex.com*

[32] "A German Diesel-Pneumatic Locomotive" (http://www.douglas-self.com/MUSEUM/LOCOLOCO/diesair/diesair.htm). Douglas-self.com. . Retrieved 2011-08-20.

[33] (http://www.strps.org.uk/str/stocklist/locos/04-Naworth.htm)

[34] "Development of the Flameproof Diesel Vehicle Applied New Exhaust Gas Dry Type Treatment System" (http://sciencelinks.jp/j-east/article/200210/000020021002A0274001.php/). Sciencelinks.jp. 2009-03-18. . Retrieved 2011-08-20.

[35] King, Joe (2008-09-22). "Engineering gets $1 million grant to make locomotives leaner, greener" (http://www.niu.edu/northerntoday/2008/sept22/locomotive.shtml). Northern Illinois University. . Retrieved 2011-08-06.

[36] Eilperin, Juliet (2006-08-14). "Attention to Locomotives' Emissions Renewed" (http://www.washingtonpost.com/wp-dyn/content/article/2006/08/13/AR2006081300530.html). *Washington Post*. . Retrieved 2011-08-06.

[37] Hawthorne, Michael (February 14, 2011). "Metra finds 'alarming' pollution on some trains" (http://articles.chicagotribune.com/2011-02-14/news/ct-met-metra-air-testing-20110213_1_scott-fruin-outbound-trains-oldest-trains). *Chicago Tribune*. . Retrieved 2011-08-06.

[38] Wilkins, Davell (2011-04-13). "Study: Installed Traps In Diesel Engines Reduce Pollution Levels" (http://topnews.us/content/238438-study-installed-traps-diesel-engines-reduce-pollution-levels). *Top News*. . Retrieved 2011-08-06.

[39] "Pollution on Metra Trains Worse Than Thought: Report" (http://www.myfoxchicago.com/dpp/news/metro/metra-train-cars-pollution-soot-worse-report-20110214). *Fox Chicago News*. 2011-02-14. . Retrieved 2011-08-06.

[40] Lydersen, Kari (April 21, 2011). "Black Carbon Testing Finds High Levels" (http://www.nytimes.com/2011/04/22/us/22cnccarbon.html). *The New York Times*. . Retrieved August 6, 2011.

External links

- US Government test of GP38-2 locomotive with biodiesel fuel. (http://www.nrel.gov/docs/fy04osti/33436.pdf)
- A 1926 article *The Diesel Engine in Railway Transportation* on Diesel locomotives (http://www.nzetc.org/tm/scholarly/tei-Gov01_04Rail-t1-body-d9.html)
- Diesel locomotive (http://www.railway-technical.com/diesel.shtml)

Oceanside,_California

Oceanside

— City —

An Oceanside beach

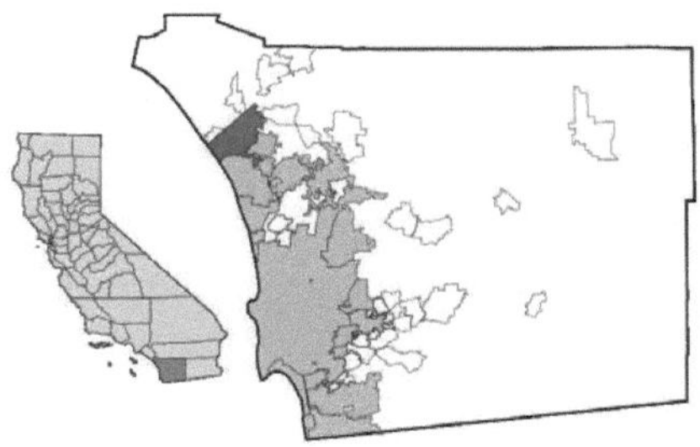

Location of Oceanside within San Diego County, California

Oceanside

Location in the United States

Coordinates: 33°12′42″N 117°19′33″W

Country	United States
State	California
County	San Diego
Government	

• Type	Council-Manager
• City Council	Jim Wood (Mayor) Jack Feller Jerome M. Kern Esther C. Sanchez Vacant Seat
• City Treasurer	Michele Lund, CCMT
• City Clerk	Barbara Riegel Wayne
• City Manager	Peter Weiss
Area[1]	
• Total	42.174 sq mi (109.231 km^2)
• Land	41.235 sq mi (106.798 km^2)
• Water	0.939 sq mi (2.433 km^2) 2.23%
Elevation	66 ft (20 m)
Population (2010 Census)	
• Total	183,095
• Rank	3rd in San Diego County 27th in California 140th in the United States
• Density	/sq mi (**unknown operator: u',**'/km^2)
Time zone	PST (UTC-8)
• Summer (DST)	PDT (UTC-7)
ZIP code	92049, 92051-92052, 92054, 92056-92058
Area code(s)	760, 442
FIPS code	06-53322
GNIS feature ID	1652761
Website	http://www.ci.oceanside.ca.us/

Oceanside is a coastal resort city located on California's South Coast. It is the third-largest city in San Diego County, California. The city had a population of 183,095 at the 2010 census. Together with Carlsbad and Vista, it forms a tri-city area. Oceanside is located just south of Marine Corps Base Camp Pendleton, the busiest military base in the United States.[2] The city has experienced dramatic growth since 1970, when its population was 45,000. Much of the city's area was developed into single-family home tracts during the 1970s and 1980s. Since the 1990s, increased commercial and industrial development have diversified Oceanside's economic base.

History

Andrew Jackson Myers, Oceanside's founder

Although the area was first settled by Native Americans, the first European explorers arrived in 1769. The Spanish missionaries under Father Junipero Serra founded Mission San Luis Rey de Francia on a former site of a Luiseño Indian village on the banks of the San Luis Rey River. In the early 19th century, the introduction of farming and grazing changed the landscape of what would become Oceanside. The area - like all of California - was under Spanish, then in 1821 under Mexican rule, and conquered by the U.S. in 1848.

In the late 1850s, Andrew Jackson Myers lived in San Joaquin County. A native of LaSalle County, Illinois, he returned in the late 1880s and lived in San Luis Rey. In 1882 Myers moved on the land that was the original town site for Oceanside. A patent for the land was issued in 1883 by the federal government.[3] It was incorporated on July 3, 1888. The city hall as of the early 21st century stands on the former Myers homestead.[3]

In the 20th century, Oceanside was a beach town devoted to activities on a 6-mile (9.7 km) stretch of beaches. Residential areas like downtown (built in the 1890s), South Oceanside (built in the 1920s and 1930s), and developments east of Interstate 5 (built after World War II) are preserved and remodeled when these houses are considered to have historical value. Since the establishment of Marine Corps Base Camp Pendleton in 1942, Oceanside has been home to U.S. armed forces personnel, and the wartime industry of WWII and the 1950s had an ammunition manufacturing facility in the city. After 1970, the main focus of Oceanside was suburban development and a choice for newcomers to move in to then relatively affordable housing.

Geography

Oceanside is at 33°12′42″N 117°19′33″W (33.211566, -117.325701).[4]

According to the United States Census Bureau, the city has a total area of 42.2 square miles (109 km^2). 41.2 square miles (107 km^2) of it is land and 0.9 square miles (2.3 km^2) of it (2.23%) is water.

Climate

Oceanside experiences a semi-arid climate (Köppen climate classification *BSh*).

Demographics

2010

Oceanside City Hall complex

The 2010 United States Census[5] reported that Oceanside had a population of 167,086. The population density was 3,961.8 people per square mile (1,529.7/km²). The racial makeup of Oceanside was 109,020 (65.2%) White, 7,873 (4.7%) African American, 1,385 (0.8%) Native American, 11,081 (6.6%) Asian (3.4% Filipino, 0.7% Japanese, 0.7% Vietnamese, 0.6% Chinese, 0.4% Korean, 0.2% Indian), 2,144 (1.3%) Pacific Islander, 25,886 (15.5%) from other races, and 9,697 (5.8%) from two or more races. Hispanic or Latino of any race were 59,947 persons (35.9%). Among the Hispanic population, 31.3% Mexican, 1.0% Puerto Rican, and 0.4% Salvadoran.

The Census reported that 166,150 people (99.4% of the population) lived in households, 802 (0.5%) lived in non-institutionalized group quarters, and 134 (0.1%) were institutionalized.

There were 59,238 households, out of which 20,486 (34.6%) had children under the age of 18 living in them, 30,201 (51.0%) were opposite-sex married couples living together, 6,947 (11.7%) had a female householder with no husband present, 3,111 (5.3%) had a male householder with no wife present. There were 3,504 (5.9%) unmarried opposite-sex partnerships, and 472 (0.8%) same-sex married couples or partnerships. 14,117 households (23.8%) were made up of individuals and 6,161 (10.4%) had someone living alone who was 65 years of age or older. The average household size was 2.80. There were 40,259 families (68.0% of all households); the average family size was 3.32.

The population was spread out with 39,817 people (23.8%) under the age of 18, 19,028 people (11.4%) aged 18 to 24, 45,797 people (27.4%) aged 25 to 44, 40,943 people (24.5%) aged 45 to 64, and 21,501 people (12.9%) who were 65 years of age or older. The median age was 35.2 years. For every 100 females there were 97.4 males. For every 100 females age 18 and over, there were 95.0 males.

There were 64,435 housing units at an average density of 1,527.8 per square mile (589.9/km²), of which 34,986 (59.1%) were owner-occupied, and 24,252 (40.9%) were occupied by renters. The homeowner vacancy rate was 2.2%; the rental vacancy rate was 6.2%. 97,645 people (58.4% of the population) lived in owner-occupied housing units and 68,505 people (41.0%) lived in rental housing units.

2000

As of the census[6] of 2000, there were 161,029 people, 56,488 households, and 39,259 families residing in the city. The population density was 3,967.2 inhabitants per square mile (1,531.7/km²). There were 59,581 housing units at an average density of 1,467.9 per square mile (566.7/km²). The racial makeup of the city was 53.6% White, 30.2% Hispanic, 6.3% African American, 5.5% Asian, 1.2% Pacific Islander, 0.4% Native American or Al tive, 0.1% from another race alone, and 3.2% from two or more races. (These figures have been adjusted to classify Hispanics as a separate group from whites, blacks, Asians, and other races; U.S. census data do not separate out Latinos in this manner.)

The area around Oceanside High School was, along with the neighborhood around San Diego High School, the site of the first Samoan communities on the U.S. mainland.

In 2000, there were 56,488 households out of which 35.0% had children under the living with them, 54.1% were married couples living together, 11.0% had a female householder with no husband present, and 30.5% were non-families. 22.7% of all households were made up of individuals and 10.2% had someone living alone who was 65 years of age or older. The average household size was 2.83 and the average family size was 3.33.

Looking South over Oceanside Blvd.

The age distribution of Oceanside in 2000 was as follows: 27.6% under the age of 18, 10.2% from 18 to 24, 31.0% from 25 to 44, 17.6% from 45 to 64, and 13.6% who were 65 years of age or older. The median age was 33 years. For every 100 females there were 98.0 males. For every 100 females age 18 and over, there were 96.1 males.

The median income for a household in the city was $46,301, and the median income for a family was $52,232. Males had a median income of $34,772 versus $27,962 for females. The per capita income for the city was $20,329. About 8.2% of families and 11.6% of the population were below the poverty line, including 16.2% of those under age 18 and 5.7% of those age 65 or over.

According to estimates by the San Diego Association of Governments, the median household income of Oceanside in 2005 was $61,792 (not adjusted for inflation). When adjusted for inflation, the median household income was $50,177. On June 13, 2005, the Wall Street Journal rated Oceanside as the top vacation home market in the country.

Government

Local government

According to the city's most recent Comprehensive Annual Financial Report, the city's various funds had $295.2 million in revenues, $252.8 million in expenditures, $962.8 million in total assets, $195.3 million in total liabilities, and $223.7 million in cash and investments.[7]

The structure of the management and coordination of city services is:[8]

City department	Director
City Manager	Peter Weiss
City Attorney	John P. Mullen
City Clerk	Barbara Riegel Wayne
City Treasurer	Gary Felien
Deputy City Manager	Michelle Skaggs Lawrence
Deputy City Manager	Donald L. Hadley
Economic and Community Development Director	Jane D. McVey
Financial Services Director	Teri Ferro
Fire Chief	Darryl Hebert
Harbor and Beaches Manager	Ray Duncan
Housing & Neighborhood Services Director	Margery Pierce
Library Director	Deborah Polich
Human Resources Director	Brian Kammerer
Police Chief	Frank McCoy
Deputy Public Works Director	Joseph Arranaga
Water Utilities Director	Lonnie Thibodeaux
Interim Development Services Director	Lauren Wasserman
Chief Information Technologies Officer	Michael Sherwood

Politics

In the state legislature Oceanside is located in the 38th Senate District, represented by Republican Mark Wyland, and in the 73rd and 74th Assembly District, represented by Republicans Diane Harkey and Martin Garrick respectively. Federally, Oceanside is located in California's 49th congressional district, which has a Cook PVI of R +10[9] and is represented by Republican Darrell Issa.

Schools

Residents of Oceanside may attend schools in the Oceanside Unified School District, Vista Unified School District, Bonsall Union School District, or Carlsbad Unified School District, depending on their actual address. The Oceanside Unified School District provides instrumental music programs in grades 4-12 as well as free transportation for students in grades K-5, with the exception for those who reside on Camp Pendleton and the special needs students.

The Oceanside Unified School District has two comprehensive high schools, El Camino High School off Rancho Del Oro and Oceanside High School off Mission Avenue. High school students are also served by Ocean Shores Continuation High School and Clair Burgener Academy. OUSD has 24 schools plus three charter schools, including the School of Business and Technology, and two brand new schools, Louise Foussat Elementary School and Cesar E. Chavez Middle School, that opened in the Fall of 2007. Cesar Chavez Middle School, which is on the corner of Frazee and Oleander, will be built on a 14 acres (57000 m^2), house 11 building totaling 84000 square feet (7800 m^2) and will serve 1,000 6th - 8th grade students.

The other school, Louise Foussat Elementary School, located on Pala Road, is built on 12.6 acres (51000 m^2) of land with 35 classrooms totaling 54, 490 square feet (46 m^2) and can accommodate 800 students.

Oceanside also serves as the home of The Teaching, Education and Research Institute (TERI Inc), a multifaceted non profit organization which serves over 600 individuals with cognitive or intellectual disabilities and employees over 400 full time and part time employees.

Economy

The headquarters for Genica Corporation (Geeks.com), Bergensons Property Services, Learning Forum International and International Stem Cell Corporation are among some of the companies located in Oceanside. According to the City's 2009 Comprehensive Annual Financial Report,[10] the top employers in the city are:

Oceanside Marina

#	Employer	# of Employees
1	Tri-City Medical Center	1,900
2	City of Oceanside	1,060
3	Mira Costa College	1,000
4	Select Staffing	600
5	North County Transit District	600
6	Genentech	500
7	Milagro Ranch	400
8	Ocean's Eleven Casino	400
9	Hydranautics	300
10	Registry Network	300

Religion

The city of Oceanside is uniquely located in the center of three religious philosophies. The spiritual triangle of Oceanside features the Prince of Peace Benedictine Abbey to the north, the Rosicrucian Fellowship to the west, and the Mission San Luis Rey de Francia, a Franciscan Mission to the east.

Attractions

Looking south from the Oceanside Pier

- The Oceanside Pier, first built in 1888 (and now in its sixth incarnation), is the longest wooden pier on the western United States coastline at 1,954 feet (596m).
- Oceanside is home to Mission San Luis Rey de Francia, one of the Alta California missions.
- The historic district of Mount Ecclesia, home to the Rosicrucian Fellowship, is noted for its singular architecture and the preservation of nature grounds and gardens, offering a unique meditative walking experience.
- The bungalow house featured in *Top Gun*, starring Tom Cruise, is located on South Pacific Street, which is one street up from the Oceanside Strand, a section of residential houses along the coastline.[11]
- In June 1997, a record 221,000 fans watched X Games III, held in San Diego and Oceanside. The surfing competition took place at the Oceanside Harbor Beach.
- The California Surf Museum is located in downtown Oceanside.

Oceanside Pier

- Oceanside has a large Day of the Dead celebration held every year on or about the weekend nearest November 1 with Carnival rides, street vendors, and musicians
- Usually held the second weekend in September, Oceanside also holds an annual Oceanside Harbor Days festival with street vendors, food, and local artists displaying their wares. More information can be found at http://www.OceansideHarborDays.com
- Oceanside is host to the world's 2nd largest Beach Soccer festival, The Southern California Beach Soccer Championships has become an international event and it is usually held in the middle of May at the Oceanside Harbor Beach.
- The final competition scene from the film *Bring It On* was not shot in Daytona, Florida, but at the beach shell in front of the pier.
- In the late 1990s, Oceanside had a minor league baseball franchise of the Western Baseball League, in the Miracosta College Ballpark but folded operations and lacked a standard ball park, now the site of the North County Waves semi-pro collegiate team.
- The hills of Oceanside are said to form a lying down person in the shape of the "Sleeping Indian" near Papagallo Rd. in Oceanside.
- The 2009 film *To Save A Life* was filmed in Oceanside.

Sister cities

- Ensenada, Mexico
- Kisarazu, Japan
- Fuji, Japan
- A'ana, Samoa

Notable natives and residents

- Bobbi DePorter - Founder of the internationally acclaimed SuperCamp program, and President of the Quantum Learning Network.[12]
- Bill Goldberg (wrestler) - Retired professional wrestler, football player, and actor; see Goldberg (wrestler)[13]
- Frank Lasee, Wisconsin politician
- Rear Admiral (Ret.) Dick Lyon - One of the first 10 Navy Seals (formerly called "Scouts and Raiders"), and the first Special Warfare Officer to attain that rank. He served as mayor of the city in the 1990s.[14]
- Barbara Mandrell - Country singer. Former Miss Oceanside.[15] Graduated from Oceanside High School in 1967.[16]
- Denise Richards - Actress who starred in *Starship Troopers*, *Wild Things* and the James Bond film *The World Is Not Enough*. Graduated from El Camino High School in 1989.[17]
- Chris Thile - Virtuoso mandolinist of Nickel Creek; born in Oceanside.[18]
- Evan Tanner, former UFC middleweight champion
- Francis French, author and noted space historian
- Thomas Keller, chef, restaurateur and Michelin Star recipient.

- Victor Villaseñor, acclaimed Mexican-American writer
- Jordan Pundik, lead singer of pop punk band New Found Glory
- Christopher Lloyd, best known for his role in the Back to the Future films
- Brandon Vera, UFC Light-Heavyweight Mixed Martial Artist
- The Drowning Men, indie-rock band
- Kassia Meador, Professional female surfer, and photographer

Major League Baseball players

- Heath Bell, pitcher for the Miami Marlins
- Thad Bosley, player for the California Angels, Chicago White Sox, Milwaukee Brewers, Seattle Mariners, Chicago Cubs, Kansas City Royals, and the Texas Rangers
- Trevor Cahill, pitcher for the Oakland Athletics
- Chris Chambliss, who played from 1971 to 1988 for the Cleveland Indians, New York Yankees and Atlanta Braves, graduated from Oceanside High School in 1966

National Football League players

- Willie James Buchanon, played for the Green Bay Packers and San Diego Chargers. Graduated from Oceanside High School in 1968.
- D. J. Clark, played for the Green Bay Packers.[19]
- Joe Salave'a, plays for the Washington Redskins.[20]
- Junior Seau - linebacker who played for the San Diego Chargers for 13 years, before being traded to the Miami Dolphins in 2003 and most recently the New England Patriots. Graduated from Oceanside High School with letters in football, basketball, and track.[21]
- Ken Stills, played for the Green Bay Packers and Minnesota Vikings.[22]
- Toussaint Tyler, played running back for the New Orleans Saints for two seasons[23]

Films

- *Battle: Los Angeles*
- *Bring It On*
- *Top Gun*[11]
- *The Whole Ten Yards*
- *Veronica Mars*
- *Heartbreak Ridge*
- *Diamonds Are Forever*
- *Attack of the Killer Tomatoes*
- *The Mentalist*
- *Captain America 1990*
- *To Save a Life*

References

[1] U.S. Census (http://www.census.gov/geo/www/gazetteer/files/Gaz_places_national.txt)
[2] Marine Corps Base (http://www.cpp.usmc.mil/cpao/pages/baseinformation.asp) Camp Pendleton
[3] "Image:Oceanside plaque" (http://commons.wikimedia.org/wiki/Image:Oceanside_plaque.jpg). . Retrieved July 17, 2006.
[4] "US Gazetteer files: 2010, 2000, and 1990" (http://www.census.gov/geo/www/gazetteer/gazette.html). United States Census Bureau. 2011-02-12. . Retrieved 2011-04-23.
[5] All data are derived from the United States Census Bureau reports from the 2010 United States Census, and are accessible on-line here (http://factfinder2.census.gov). The data on unmarried partnerships and same-sex married couples are from the Census report DEC_10_SF1_PCT15. All other housing and population data are from Census report DEC_10_DP_DPDP1. Both reports are viewable online or downloadable in a zip file containing a comma-delimited data file. The area data, from which densities are calculated, are available on-line here (http://www.census.gov/geo/www/gazetteer/files/Gaz_places_national.txt).

Percentage totals may not add to 100% due to rounding. The Census Bureau defines families as a household containing one or more people related to the householder by birth, opposite-sex marriage, or adoption. People living in group quarters are tabulated by the Census Bureau as neither owners nor renters. For further details, see the text files accompanying the data files containing the Census reports mentioned above.

[6] "American FactFinder" (http://factfinder.census.gov). United States Census Bureau. . Retrieved 2008-01-31.
[7] City of Oceanside CAFR (http://www.ci.oceanside.ca.us/pdf/CAFR2008_financial.pdf) Retrieved 2009-08-13
[8] City of Oceanside CAFR (http://www.ci.oceanside.ca.us/pdf/CAFR2008_intro.pdf) Retrieved 2009-08-13
[9] "Will Gerrymandered Districts Stem the Wave of Voter Unrest?" (http://www.clcblog.org/blog_item-85.html). Campaign Legal Center Blog. . Retrieved 2008-02-10.
[10] City of Oceanside CAFR (http://www.ci.oceanside.ca.us/pdf/CAFR2009_statistical.pdf) Retrieved 2009-08-13
[11] McIntosh, Linda (May 22, 2011). "Cruise into movie history at "top gun house"" (http://www.signonsandiego.com/news/2011/may/22/cruise-movie-history-top-gun-house/). *The San Diego Union-Tribune*. . Retrieved May 25, 2011.
[12] "Q&A: Bobbi DePorter; founder of Quantum Learning Network" (http://sports.uniontrib.com/uniontrib/20060514/news_m1m14qa.html). The San Diego Union-Tribune. . Retrieved 2009-05-10.
[13] The Hebrew Hulk (http://www.sdjewishjournal.com/stories/cover_may05.html)
[14] "Elected officials are skilled — at getting elected" (http://www.nctimes.com/articles/2008/01/25/opinion/rider/20_51_591_24_08.txt). North Country Times. . Retrieved 2009-05-10.
[15] The International Who's Who of Women 2002 (http://books.google.com/books?id=6J8xDWDqOkEC&pg=RA2-PA359&lpg=RA2-PA359&dq="barbara+mandrell"+"miss+oceanside"&source=web&ots=vMBS-MdEBT&sig=_dAp8fOzLpJaiiBMjTq4S-8th3Y)
[16] Barbara Mandrell (http://www.nndb.com/people/932/000024860/)
[17] Denis Richards Pictures (http://www.moono.com/html/denise-richards/denise-richards-pictures.cfm)
[18] Edgar Meyer and Chris Thile (http://www.aetn.org/programs/aetnpresents/archives/news_items/edgar_meyer_and_chris_thile)
[19] D.J. Clark (http://www.nfl.com/players/d.j.clark/profile?id=CLA263421)
[20] Joe Salave'a (http://www.joesalavea.com/joenfl.asp)
[21] Junior Seau Foundation (http://www.juniorseau.org/about_us_junior.htm)
[22] http://www.pro-football-reference.com/players/S/StilKe20.htm
[23] " Toussaint Tyler NFL & AFL Football Statistics (http://www.pro-football-reference.com/players/T/TyleTo00.htm)". *Pro-Football Reference*. Retrieved February 10, 2010.

External links

- Official City of Oceanside website (http://www.ci.oceanside.ca.us/)
- Oceanside travel guide from Wikitravel
- Mission San Luis Rey de Francia (http://missiontour.org/sanluisrey/index.htm)

Escondido,_California

Escondido

— City —

City of Escondido

Downtown — Grand Avenue.

Nickname(s): Hidden Valley; The Hidden City; The Heart of San Diego North; Esco

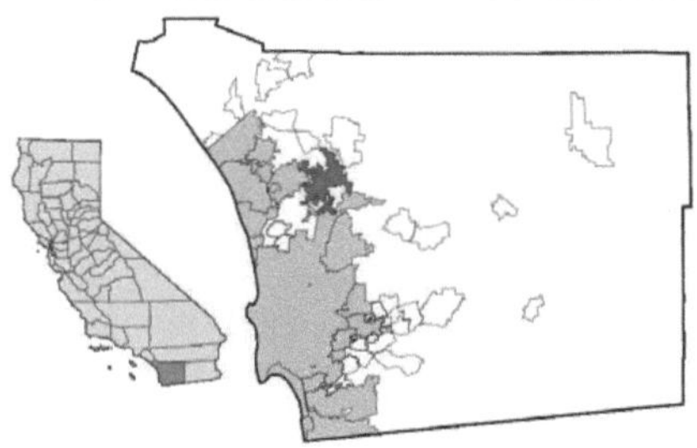

Location of Escondido, California

Coordinates: 33°7′29″N 117°4′51″W

Country	United States
State	California
County	San Diego
Incorporated	October 8, 1888
Government	
• Type	Council-Manager
• Mayor	Sam Abed
Area[1]	
• Total	36.989 sq mi (95.801 km^2)

• Land	36.813 sq mi (95.345 km^2)
• Water	0.176 sq mi (0.456 km^2) 0.48%
Elevation	646 ft (197 m)
Population (2010)	
• Total	143911
• Rank	4th in San Diego County 38th in California 171st in the United States
• Density	3890.6/sq mi (1502.2/km^2)
Time zone	PST (UTC-8)
• Summer (DST)	PDT (UTC-7)
ZIP Code	92025-92027, 92029
Area code(s)	760 and 442 (overlay plan)
FIPS code	06-22804
GNIS feature ID	1652706
Website	www.escondido.org [2]

Escondido (/ˌɛskənˈdiːdoʊ/ ***es-kən-dee-doh***; Spanish for "hidden") is a city occupying a shallow valley ringed by rocky hills, just north of the city of San Diego, California. Founded in 1888, it is one of the oldest cities in San Diego County. The city had a population of 143,911 at the 2010 census. Its municipal government set itself an operating budget limit of $426,289,048 for the fiscal year 2010-2011.[3] The city is known as *Eskondiid* in Diegueño.[4] A 2005 nationwide study of the most conservative cities in America ranked Escondido #11 out of 25.[5]

History

The Escondido area was first settled by the Luiseño, who established campsites and villages along the creek running through the area. They named the place "Mehel-om-pom-pavo." The Kumeyaay migrated from areas near the Colorado River, settling both in the San Pasqual Valley and near the San Dieguito River in the southwestern and western portions of what is now Escondido. Most of the villages and campsites today have been destroyed by development and agriculture.[6]

Spain controlled the land from the late 18th century to the early 19th century, and established many missions in California to convert the indigenous people. When Mexico gained its independence from Spain, the local land was divided into large ranchos. Most of what is now Escondido occupies the former Rancho Rincon del Diablo("Devil's Corner"), a Mexican land grant given to Juan Bautista Alvarado (not the governor of the same name)in 1843 by Governor Manuel Micheltorena. Alvarado was a Regidor of Los Angeles at the time, and the first Regidor of the pueblo of San Diego. The southern part of Escondido occupies the former Rancho San Bernardo, granted in 1842 and 1845.[7]

In 1846, during the Mexican-American War, the Battle of San Pasqual was fought southeast of Escondido. This battle pitted Mexican forces under Andrés Pico (brother of then-California-governor Pío Pico) against Americans under Stephen W. Kearny, Archibald Gillespie, and Kit Carson. A park in Escondido is named for Carson.

In 1853, pro-Southern Copperheads proposed dividing the state of California to create a new Territory of Colorado (at this time the territory that would become the state of Colorado was named "Jefferson"). San Diego Judge Oliver S. Witherby suggested placing the capitol of the new territory in Rancho Rincon del Diablo. He envisioned a railroad connecting San Diego to Fort Yuma through an area about two miles (3 km) south of the current Escondido site, heading east through San Pasqual. He planned to profit from the town that he believed would be established from the dividing point on the railroad below the eastern hills.[8]

With a series of deeds in 1855 and 1856, the rancho was transferred from the heirs of Juan Bautista Alvarado to Witherby.[7]

The proposal for splitting the state and creating the new territory passed in the California legislature, but died in Congress in the run-up to the Civil War.[7] It was effectively killed in 1861 when Congress organized the Territory of Colorado in the area previously occupied by the Jefferson Territory. With Witherby's vision of owning a bustling state capitol unrealized, he set up a mining operation on the rancho instead.[8]

In 1868, Witherby sold the rancho for $8000 to Edward McGeary and John, Josiah, and Matthew Wolfskill. McGeary owned half the rancho, while the three Wolfskill brothers each owned an equal share of the other half. John Wolfskill farmed sheep, horses, and cattle on the rancho for a number of years. Wolfskill had frequent conflicts with the Couts family, owners of the neighboring Guajome, Buena Vista, and San Marcos ranchos, over grazing lands and watering holes.[7]

Incorporation and growth

The city was home to a large Spanish-speaking population in the first census, in 1850, but after the U.S. won the war, non-Hispanic settlers came to Southern California in increasing numbers. The decade of the 1880s is known as the "Southern California Land Boom" because so many people moved to the state.

In October 1883, a group of Los Angeles investors purchased Rancho Rincon del Diablo. This group sold the land to the newly formed Escondido Company in 1884. On December 18, 1885, investors incorporated the Escondido Land and Town Company, and in 1886 this company purchased the 12814-acre (52 km^2) area for approximately $100,000.[7] Two years later, in 1888, Escondido was incorporated as a city; the vote was 64 in favor of cityhood with 12 votes against. Railroads such as the Santa Fe and Southern Pacific were laid in the 1880s. The opening of U.S. Route 395 in 1930 boosted economic growth in Escondido.

Escondido was primarily an agricultural community, growing muscat grapes initially. After a dam was built in 1894-5 to form what is known today as Lake Wohlford, orange and lemon trees were planted in large numbers, as were olive and walnut trees. By the 1960s avocados became the largest local crop. Since the 1970s, Escondido has lost most of its agricultural land to housing developments.

Geography

Escondido is located at 33°7'29" North, 117°4'51" West (33.124794, -117.080850 [9]).[10]

Looking south west across Escondido from the hills near Dixon Lake.

According to the United States Census Bureau, the city has a total area of 37.0 square miles (96 km^2). 36.8 square miles (95 km^2) of it is land and 0.2 square miles (0.52 km^2) of it is water. The total area is 0.48% water.

The city is growing at a rapid rate with new communities like Hidden Trails [11] appearing at the east end of East Valley Parkway. The city proper is surrounded by several sparsely populated unincorporated communities. These include Jesmond Dene and Hidden Meadows to the north; Felicita Park to the southwest; and Rincon Del Diablo to the southeast. Residents of these communities have Escondido mailing addresses and zip codes, and their children are sometimes assigned to Escondido schools, but residents of these communities cannot participate in city elections.

The city contains several neighborhoods including:

- Downtown Escondido centers around Grand Avenue between Centre City Parkway and Palomar Hospital. The city's general plan defines the Downtown Specific Plan Area as approximately 460 acres (1.9 km^2) bounded by Centre City Parkway on the west, Hickory and Ivy Streets on the east, Washington Avenue on the north, and Fifth Avenue on the south, with an additional narrow section extending west along Valley Parkway to Interstate 15.[12]
- Old Escondido is bounded by Escondido Boulevard on the west, Chestnut Street on the east, Fifth Avenue on the north, and Thirteenth Avenue on the south. This area is made up of mostly single-family residential housing.[13]

The community of Valley Center is located just north of Escondido. Valley View Casino, owned by the San Pasqual Band of Diegueno Mission Indians, is located in Valley Center.

Climate

Escondido tends to have a typical Mediterranean climate with warm summers and cool wet winters. Owing to its inland proximity it is considerably warmer than coastal cities like San Diego, Carlsbad or Oceanside. Yearly precipitation averages around 15 inches (380 mm) and can vary considerably from year to year. More than 80% of all precipitation takes place from November through March. Snow is virtually unheard of though occasionally Springtime thunderstorms will drop small hail. The climate is mild enough to allow widespread cultivation of avocados and oranges. Escondido is located in a plant hardiness zone 9.

Lakes

Three lakes are located in or near Escondido, all of which allow boating and fishing:

- Dixon Lake is located in the north.[15]
- Lake Wohlford is located on the north eastern fringe of the city limits.[16]
- Lake Hodges is located just south of the city. A part of Lake Hodges is crossed by Interstate 15 via the Lake Hodges Bridge. Often this part of the lake is dry.

Demographics

Historical populations		
Census	Pop.	%±
1890	541	—
1900	755	39.6%
1910	1334	76.7%
1920	1789	34.1%
1930	3421	91.2%
1940	4560	33.3%
1950	6544	43.5%
1960	16377	150.3%
1970	36792	124.7%
1980	64355	74.9%
1990	108635	68.8%
2000	133559	22.9%
2010	143911	7.8%

2010 census

In the 2010 United States Census,[17] Escondido had a population of 143,911. The population density was 3,890.7 people per square mile (1,502.2/km²). The racial makeup of Escondido was 60.4% White (40.4% non-Hispanic white), 2.5% African American (2.1% non-Hispanic black), 1.0% Native American, 6.1% Asian, 0.2% Pacific Islander, 25.4% from other races, and 4.4% from two or more races. Hispanic or Latino of any race were 48.9% of the population.

The Census reported that 141,792 people (98.5% of the population) lived in households, 1,333 (0.9%) lived in non-institutionalized group quarters, and 786 (0.5%) were institutionalized.

There were 45,484 households, out of which 18,989 (41.7%) had children under the age of 18 living in them, 23,535 (51.7%) were opposite-sex married couples living together, 6,082 (13.4%) had a female householder with no husband present, 3,115 (6.8%) had a male householder with no wife present. There were 3,121 (6.9%) unmarried opposite-sex partnerships, and 343 (0.8%) same-sex married couples or partnerships. 9,528 households (20.9%) were made up of individuals and 4,235 (9.3%) had someone living alone who was 65 years of age or older. The average household size was 3.12. There were 32,732 families (72.0% of all households); the average family size was 3.57.

The population was spread out with 39,778 people (27.6%) under the age of 18, 15,455 people (10.7%) aged 18 to 24, 41,043 people (28.5%) aged 25 to 44, 32,551 people (22.6%) aged 45 to 64, and 15,084 people (10.5%) who were 65 years of age or older. The median age was 32.5 years. For every 100 females there were 98.2 males. For every 100 females age 18 and over, there were 96.1 males.

There were 48,044 housing units at an average density of 1,298.9 per square mile (501.5/km²), of which 23,759 (52.2%) were owner-occupied, and 21,725 (47.8%) were occupied by renters. The homeowner vacancy rate was 2.2%; the rental vacancy rate was 6.0%. 70,936 people (49.3% of the population) lived in owner-occupied housing units and 70,856 people (49.2%) lived in rental housing units.

2000 census

As of the census[18] of 2000, there are 133,559 people, 43,817 households, and 31,153 families residing in the city. The population density is 1,421.4/km² (3,680.9/mi²). There are 45,050 housing units at an average density of 479.4/km² (1,241.6/mi²). The racial makeup of the city is 67.82% White, 2.25% African American, 1.23% Native American, 4.46% Asian, 0.23% Pacific Islander, 19.19% from other races, and 4.81% from two or more races. 38.70% of the population are Hispanic or Latino of any race.

There are 43,817 households out of which 39.1% have children under the age of 18 living with them, 53.9% are married couples living together, 11.7% have a female householder with no husband present, and 28.9% are non-families. 22.4% of all households are made up of individuals and 10.1% have someone living alone who is 65 years of age or older. The average household size is 3.01 and the average family size is 3.50.

In the city the population is spread out with 29.7% under the age of 18, 10.4% from 18 to 24, 31.4% from 25 to 44, 17.5% from 45 to 64, and 11.0% who are 65 years of age or older. The median age is 31 years. For every 100 females there are 98.4 males. For every 100 females age 18 and over, there are 96.1 males.

The median income for a household in the city is $42,567, and the median income for a family is $48,456. Males have a median income of $32,627 versus $27,526 for females. The per capita income for the city is $18,241. 13.4% of the population and 9.3% of families are below the poverty line. Out of the total population, 17.9% of those under the age of 18 and 5.7% of those 65 and older are living below the poverty line.

The city can be divided into two demographically distinct areas. Peripheral hilly areas to the north, southeast, and southwest are relatively wealthy and populated by non-Hispanic whites, and flat areas adjacent to the downtown are predominantly Hispanic. As of 2006-07 school year, non-Hispanic white children comprised 71.7% of all students in Bernardo Elementary School (southwest), 60.8% of all students in L.R. Green Elementary School (southeast), and 54.7% of all students in Reidy Creek Elementary School (north); on the other hand, Farr Avenue, Pioneer and

Lincoln Elementary schools (three large schools just north of the downtown) all have more than 85% of Hispanic and less than 6% non-Hispanic white students.

Crime

In 2007, the city ranked #65 by violent crimes per capita and #58 by property crimes per capita among 165 cities in California with populations greater than 50,000. Compared with the 12 largest cities in San Diego County, it ranked 6th in both categories. Its crime rate was lower in both categories than San Diego, El Cajon, and National City; higher in both categories than San Marcos, Carlsbad, Encinitas, and Santee. Escondido had a higher violent crime rate but lower property crime rate than La Mesa and Chula Vista; it had a lower violent crime rate but higher property crime rate than Vista and Oceanside.

In 2009, 629 violent crimes and 3,880 property crimes were reported in Escondido. There were four murders and non-negligent manslaughters, 42 rapes, 249 robberies, 334 aggravated assaults, 779 burglaries, 2,402 larceny thefts, 699 vehicle thefts, and 23 arsons.[19]

In 2010, Escondido saw a 5 percent drop in violent crime, with only 597 reported violent crimes according to the Escondido Police Officer's Association. However, there was a 3.9 percent increase in the number of property crimes, including residential and commercial burglaries, from 3,880 in 2009 to 4,033 in 2010, according to FBI statistics.[20]

Economy

Residents work in a range of industries. Out of the approximately 64,000 employed civilian residents over the age of 16, 15% work in educational, health care and social services; 13% in retail trade; 13% in construction; 12% in professional, scientific, management, administrative, and waste management services; 11% in arts, entertainment, recreation, and accommodation and food services; 11% in manufacturing; and 11% in other services.[21]

Realty Income is among the companies based in Escondido.

Top employers

According to the City's 2009 Comprehensive Annual Financial Report,[22] the top employers in the city are:

#	Employer	# of Employees
1	Palomar Medical Center	2,664
2	Escondido Union School District	1,795
3	City of Escondido	1,075
4	Escondido Union High School District	715
5	North County Transit District	484
6	ARS National Services	439
7	Nordstrom	418
8	Ne-Mo's Bakery	352
9	Palomar College	340
10	The Home Depot	333

Arts and culture

Downtown

Downtown has become more active in the past few years with the opening of restaurants, cafes, and galleries. Every Friday night from April through September, Steve Waldron and a handful of friends host the popular "Cruisin' Grand [23]", where the public can show and view hot rods and historic cars. A different car club and/or featured attraction (for example, antique fire trucks, nitro night, midget and sprint cars) is highlighted each week. Cruisin' Grand also features a DJ, hula hoop contests for children, and the awarding of trophies.

Central Grand Ave, Escondido. Photo taken June 13, 2010.

In addition to the many art galleries on Grand, a branch of the Mingei International Museum recently opened there.(now recently closed) This museum displays handcrafts from around the world. One block off Grand Ave. is Grape Day Park with the civic center and the California Center for the Arts, which features two theaters, a visual arts museum, an educational complex, and a conference center. The Escondido Children's Museum and the Escondido History Center, two independent non-profit museums, are located in Grape Day Park. The Children's Museum features hands-on exhibits and programs for children up to 10 years of age, with an authentically regional perspective on natural and social science. The History Center features the city's original Santa Fe Depot, first library, Victorian house, barn, and blacksmith shop. The Pioneer Room of Escondido Public Library (located in the Mathes Center building next to the Main Library) offers photographs, maps, oral histories, genealogical collections, directories and yearbooks documenting Escondido's history.[24]

Sports

From 1964–1968, the San Diego Chargers held training camp in Escondido.[25]

In 1981, Escondido National Little League became the 19th[26] team to make it to the Little League World Series from the state of California. The team was first District 31 champions, then District 8 champions.[27] They then won the Southern California Divisional Tournament at Youth Athletic Park by beating San Bernardino Civitan 3-2 in the quarterfinals, then beating Granada Hills American 5-1 in the semifinals and then beating Ladera National 7-5 in the finals to earn a trip to the Western Regional. At the Western Regional in San Bernardino, the Escondido team won four straight games to earn the trip to Williamsport.[28]

In October 2010, Merritt Paulson, owner of the AAA Portland Beavers franchise, announced that the team was being sold to the North County Baseball ownership group, led by Jeff Moorad, part-owner and CEO of the Beavers parent team, the San Diego Padres. The ownership group is in discussions to build a stadium in Escondido to become operational for the 2012 baseball season at the earliest. In December 2010, the Escondido city council voted to go ahead with the ballpark. The stadium is slated to open in April 2013.[29] [30]

The San Diego Sockers announced that the team may build an indoor soccer-specific stadium with a target date of fall, 2013.

Parks and recreation

Parks

Escondido has fifteen parks.[31]

- **Daley Ranch** is a 3058-acre (12.38 km^2) conservation area acquired in 1996 by the City of Escondido and is managed in perpetuity for the preservation of a biologically unique and diverse habitat area of regional importance. Daley Ranch has over 20 miles (32 km) of hiking, mountain biking, and equestrian trails. Six distinct trails cover most of the property. Most popular are the Boulder Loop Trail (2.4 miles) which affords views of Escondido, and the Ranch House Loop (2.5 miles), which passes two small ponds and the site of Daley's original log cabin.
- **El Norte Park** is a small "green lung" park, off El Norte Parkway in northern Escondido. It features mature trees and a kids playground.
- **Grape Day Park** is located behind the Escondido City Hall and the performing arts complex. This park was named after the annual grape day harvest, an event held on the first Saturday after Labor Day. Dogs are not allowed in Grape Day Park.[32]

Grape Day Park, Escondido with the city hall in the background. Photo taken June 13, 2010.

- **Jesmond Dene**
- **Kit Carson**: *Queen Califia's Magical Circle*, the last major international work by French artist Niki de Saint Phalle, is located in Kit Carson Park. De Saint Phalle, a colleague of Salvador Dalí and Jasper Johns, is best known for her Stravinsky Fountain, located in Paris, France. The artist chose Escondido as the site for her final work because it reminded her of Italy.
- **Dixon Lake** supplies Escondido with fresh water. Swimming is not allowed, but fishing is, and boats are available for hire. Dixon Lake has about 34 camp sites which look out over the lake, or over Escondido. While entrance to the park is free, the city charges $5.00 (in 2010) for taking a car into the park. However, ample free parking is available outside the gate.

Dixon Lake, just outside Daley Ranch.

- **Lake Wohlford** is in the north east corner of Escondido. The lake has some walking trails around the perimeter, and allows fishing.
- **Mountain View**
- **Rod McLeod**
- **Washington Park** is an urban park located on the eastern side of central Escondido. It features lighted facilities for baseball, softball, tennis and basketball, a 25-yard swimming pool and water slide, and a smaller children's pool. The park's numerous mature trees create plenty of shady areas and tables are provided for picnics.[33]
- **Westside**
- **Frances Ryan Park** is a 67 acres (m^2) complex of soccer fields, supported by a concession stand and public restrooms. The park is adjacent to Valley High School in eastern Escondido at the corner of Valley Parkway and El Norte Parkway. As of October 2010, the park features 7 soccer fields, 5 of which have lights for night time practice and play. The park is named for a former teacher at Escondido High School. Dogs are not allowed in Ryan Park.[34]

- **Grove Park** was so named to recognize the city's agricultural history. It was built about 2008 (approximate date). It is about 4.5 acres (18000 m^2) in size and is located in the Mission Park neighborhood.
- **Felicita County Park**

San Diego Zoo Safari Park

The San Diego Zoo Safari Park (also called by its former name, "Wild Animal Park") is located near Escondido, in the San Pasqual Valley. It is the sister park to the San Diego Zoo. The Safari Park shows animals in open habitats.

Stone Brewing Company

In 2006, Stone Brewing Company moved its headquarters and brewery from San Marcos, California to a new, much larger facility in the Quail Hills area of Escondido.[35]

Deer Park Buddhist Monastery

Deer Park Monastery is a Buddhist sanctuary that occupies 400 acres (1.6 km^2) in the hills north of Escondido and west of Daley Ranch. It is one of two monasteries in the United States under the direction of a well-known Zen teacher Thich Nhat Hanh. Deer Park Monastery is home to 27 Buddhist monks and nuns; it frequently hosts events and retreats that bring people from all over the western United States and from abroad.

Government

Sister cities[36]
Maebaru, Japan

Local

Escondido is governed by a mayor-council system. The city council consists of a mayor and four City Council members. Along with the City Treasurer, they are elected at large to four-year terms. The current mayor is Sam Abed. Current City Council members are Olga Diaz, Marie Waldron, Ed Gallo, and Michael Morasco.[37] [38] The current City Manager is Clay Phillips. The current City Treasurer is Kenneth Hugins.[39] The most recent election was held on November 2, 2010.[40]

Escondido City Hall, from Grape Day Park. Photo taken June 13, 2010.

Due to the public outcry and legal challenges to the aforementioned housing ordinance, and the election of Olga Diaz to the City Council, it has since ceased any overt attempts to lower the illegal immigrant population in the city (which, council member Sam Abed's estimates is as much as 35,000, or 25% of the city population in 2006[41]), and focused on "quality of life" issues instead.

Periodic police checkpoints are instituted to catch unlicensed drivers. An active area of debate is an overnight parking ordinance that would limit the number of cars each household can legally park on city streets.[42] The city is estimated to have lost as much as a quarter of its non-citizen population between 2006 and 2007. Latino activists attribute this to a perception of the city as hostile to immigrants.[43]

The City of Escondido is a member of the San Diego Association of Governments (SANDAG).[44]

State and federal

In the state legislature Escondido is located in the 38th Senate District, represented by Republican Mark Wyland, and in the 74th and 75th Assembly District, represented by Republicans Martin Garrick and Nathan Fletcher respectively. Federally, Escondido is located in California's 50th congressional district, which has a Cook PVI of R +5[45] and is represented by Republican Brian Bilbray.

In the United States presidential election of 2008, 53.3% of voters residing in incorporated Escondido voted for John McCain, 44.9% voted for Barack Obama, and 1.8% voted for one of the third-party candidates. Unincorporated areas were considerably more conservative: among voters in neighborhoods of Rincon Del Diablo, Hidden Meadows, and Valley Center, 62.3%, 65.5%, 66.9% of voters respectively cast their votes for John McCain.

Education

Escondido is served by the Escondido Union School District,[46] the Escondido Union High School District,[47] and the San Pasqual Union School District. The city has 19 elementary, five middle, and seven high schools.

Public high schools:

- Escondido High School
- Orange Glen High School
- San Pasqual High School
- Classical Academy High School
- Escondido Charter High School
- Center City High School
- Valley High School[48]

Middle schools:

- Bear Valley Middle School
- Del Dios Middle School
- Hidden Valley Middle School
- Mission Middle School
- Rincon Middle School

Elementary Schools

- Bernardo Elementary
- Central Elementary
- Classical Academy
- Coastal Academy
- Conway Elementary
- Farr Avenue Elementary
- Felicita Elementary
- Glen View Elementary
- Heritage K-8 Charter
- Juniper Elementary
- L. R. Green Elementary
- Lincoln Elementary
- Miller Elementary

- Nicolaysen Community Day
- North Broadway Elementary
- Oak Hill Elementary
- Orange Glen Elementary
- Pioneer Elementary
- Reidy Creek Elementary
- Rock Springs Elementary
- Rose Elementary

There is a wide range of API scores for Escondido schools, reflecting the demographic diversity of the city. As of 2009,[49] two elementary schools in the district scored above the 80th percentile of all schools in the state, and nine elementary schools scored below the 20th percentile.

The Escondido Public Library system consists of the Main Branch, the East Valley Branch, Pioneer Room, Computer Center, and a bookmobile.

Infrastructure

Transportation

Two highways serve Escondido: Route 78 and Interstate 15. Route 78 enters from the west as a freeway which ends at Broadway. The highway follows surface streets and leaves the city heading east into the San Pasqual Valley.

The North County Transit District (NCTD) operates local bus service, with the Escondido Transit Center serving as a hub. The transit center has connections to both the San Diego Metropolitan Transit System and the Riverside Transit Agency.

The Sprinter light rail line, operated by NCTD, links the transit center to Oceanside using the existing 22-mile (35 km)-long Escondido Branch trackage of the San Diego Northern Railroad. The rail line opened in 2008, making Escondido one of the first cities in the United States to operate Desiro-class diesel multiple units manufactured by Siemens in Germany. At the Oceanside Transit Center, the Sprinter connects to three commuter rail lines (the Coaster, the Metrolink Orange County Line, and the Metrolink Inland Empire-Orange County Line) and Amtrak's Pacific Surfliner regional rail line.

The California High Speed Rail Authority listed Escondido as a stop along the proposed high-speed rail system running from Southern to Northern California.[50] A section of the line between San Francisco and Los Angeles was approved by voters in the November 2008 elections.

Utilities

San Diego Gas & Electric is the electric utility for the city.[51] The City of Escondido Water Utilities serves most customers within the City while Rincon del Diablo Municipal Water District serves potable and recycled water to the greater Escondido valley and some portions of the incorporated city.[52]

Health care

Palomar Medical Center is a hospital located east of downtown. It is the only designated trauma center in northern San Diego County.[53] A second hospital, Palomar West, is under construction southwest of the interchange between Interstate 15 and State Route 78. It is scheduled to open in 2012.[54]

Notable natives and residents

- Anacani: featured singer with Lawrence Welk orchestra
- Tom Anderson: co-founder of MySpace[55]
- Lester Bangs: rock music critic
- Ken Block: rally car driver, and founder of DC Shoes.
- Benjamin Britten: English composer, famous for his opera "Peter Grimes"[56]
- E. H. Bronner: soapmaker
- Ray Conniff: bandleader and arranger
- Pete Coscarart: former Brooklyn Dodger
- Gavvy Cravath: right fielder for the Philadelphia Phillies
- Richard Denning: actor
- Robert Klark Graham: eugenicist and inventor, founder of the Repository for Germinal Choice ("Nobel Prize Sperm Bank")
- Robbie Peralta MMA UFC Fighter
- Brittany Hogan: former Miss California
- Randy Johnson: former Major League player with Atlanta Braves, current Special Asst to Oakland A's GM Billey Beane, Graduate of Escondido High School
- Jacque Jones: former baseball player for the Minnesota Twins
- Jovan Kirovski: forward for the L.A. Galaxy MLS team
- Elisabeth Kübler-Ross: founder and resident of the "Shanti Nilaya" (Home of Peace), retreat in Escondido in the late 1970s and early 1980s
- Shannon MacMillan: professional soccer player and graduate of San Pasqual High School
- Deborah Jeane Palfrey: the so-called D.C. Madam[57]
- Mark Redman: professional baseball player and graduate of Escondido High School
- Steve Reeves: bodybuilder[58]
- Travis Ryan: lead vocalist of Cattle Decapitation[59]
- Brett Salisbury: college football quarterback, author of *The Transform Diet*, and member of Escondido's 1981 Little League World Series team.
- Sean Salisbury: former ESPN anchorman and former NFL quarterback; graduate of Orange Glen High School
- Randy Vasquez: actor
- Eric Weddle: #32 of San Diego Chargers resides in Escondido[60]
- Harold Bell Wright: author in 1920s and 1930s
- Andrea Zittel: contemporary artist, graduate of San Pasqual High School

See also

- North County Times
- *The Road to Escondido*, an album by J.J. Cale, Colin Tappe and Eric Clapton

References

[1] U.S. Census (http://www.census.gov/geo/www/gazetteer/files/Gaz_places_national.txt)
[2] http://www.escondido.org/
[3] [escondido.org/Data/Sites/1/pdfs/Finance/OperatingBudget.pdf City of Escondido Annual Operating Budget Fiscal Year 2010-2011, Mayor Lori Holt Pfeiler, City Council and City Manager Clay Phillips]
[4] Ted Couro and Christina Hutcheson (1973). *Dictionary of Mesa Grande Diegueño*. Morongo Indian Reservation, Banning, California: Malki Museum Press
[5] Study Ranks America's Most Liberal and Conservative Cities (http://govpro.com/content/gov_imp_31439/).govpro.com.
[6] "The History of Escondido" (http://web.archive.org/web/20070315105101/http://www.library.escondido.org/pioneer/history.htm). Escondido Public Library. Archived from the original (http://www.library.escondido.org/pioneer/history.htm) on March 15, 2007. .

Retrieved August 5, 2010.
[7] Rush, Philip S. (1965). "Rincon del Diablo". *Some old ranchos and adobes*. San Diego, California. pp. 46, 47. LCCN 6521995.
[8] Stanford, Leland G. (1978). "Devil's Corner and Oliver S. Witherby" (http://www.sandiegohistory.org/journal/78spring/corner.htm). *Journal of San Diego History* (San Diego History Center) **24** (2). . Retrieved August 5, 2010.
[9] http://local.google.com/local?ll=33.124794,-117.080850&spn=0.045789,0.077419
[10] "US Gazetteer files: 2010, 2000, and 1990" (http://www.census.gov/geo/www/gazetteer/gazette.html). United States Census Bureau. 2011-02-12. . Retrieved 2011-04-23.
[11] http://www.hiddentrailshoa.com/
[12] "II. Land Use" (http://www.escondido.org/Data/Sites/1/pdfs/Planning/GP/Section2.pdf). *City of Escondido General Plan*. 1990-06-06. pp. 141, 142, 169. . Retrieved 2010-03-29.
[13] "II. Land Use" (http://www.escondido.org/Data/Sites/1/pdfs/Planning/GP/Section2.pdf). *City of Escondido General Plan*. 1990-06-06. p. 18. . Retrieved 2010-03-29.
[14] "Weatherbase: Weather for Escondido, California" (http://www.weatherbase.com/weather/weather.php3?s=268240&refer=&units=us). Weatherbase. 2011. . Retrieved on November 22, 2011.
[15] "Dixon Lake" (http://www.escondido.org/dixon-lake.aspx). City of Escondido. . Retrieved December 2, 2010.
[16] "Lake Wohlford" (http://www.escondido.org/lake-wohlford.aspx). City of Escondido. . Retrieved December 2, 2010.
[17] All data are derived from the United States Census Bureau reports from the 2010 United States Census, and are accessible on-line here (http://factfinder2.census.gov). The data on unmarried partnerships and same-sex married couples are from the Census report DEC_10_SF1_PCT15. All other housing and population data are from Census report DEC_10_DP_DPDP1. Both reports are viewable online or downloadable in a zip file containing a comma-delimited data file. The area data, from which densities are calculated, are available on-line here (http://www.census.gov/geo/www/gazetteer/files/Gaz_places_national.txt).

Percentage totals may not add to 100% due to rounding. The Census Bureau defines families as a household containing one or more people related to the householder by birth, opposite-sex marriage, or adoption. People living in group quarters are tabulated by the Census Bureau as neither owners nor renters. For further details, see the text files accompanying the data files containing the Census reports mentioned above.

[18] "American FactFinder" (http://factfinder.census.gov). United States Census Bureau. . Retrieved 2008-01-31.
[19] California - Offenses Known to Law Enforcement by State by City, 2009 (http://www.fbi.gov/ucr/cius2009/data/table_08_ca.html)
[20] "Escondido Crime" (http://www.nctimes.com/news/local/escondido/article_e459d082-2a12-517a-bb58-8897f3cd98be.html). nctimes.com. April 5, 2011. . Retrieved 2011-06-04.
[21] "C24050. Industry by occupation for the civilian employed population 16 years and older" (http://factfinder.census.gov/servlet/DTTable?_bm=y&-context=dt&-ds_name=ACS_2008_3YR_G00_&-_geoSkip=0&-CONTEXT=dt&-mt_name=ACS_2008_3YR_G2000_B24050&-tree_id=308&-_skip=0&-redoLog=false&-geo_id=16000US0622804&-search_results=01000US&-_showChild=Y&-format=&-_lang=en&-_toggle=ACS_2008_3YR_G2000_B24050). *2006-2008 American Community Survey 3-Year Estimates*. U.S. Census Bureau. . Retrieved August 1, 2010.
[22] Comprehensive Annual Financial Report - Fiscal Year Ended June 30, 2009 (http://www.escondido.org/Data/Sites/1/pdfs/Finance/CAFR2009.pdf) (Report). City of Escondido. 2009. p. 169. . Retrieved December 2, 2010.
[23] http://www.cruisingrand.com
[24] Pioneer Room (http://www.library.escondido.org/pioneer/index.htm)
[25] "Chronology 1959-1969" (http://www.chargers.com/team/history/chronology/1959-1969.html). San Diego Chargers. . Retrieved 2009-09-28.
[26] West Region State Little League Champions (http://www.unpage.org/west/wrn-state.htm)
[27] 1981 Southern California Little League Tournament Results (http://www.unpage.org/socal/cs-1981.htm)
[28] 1981 Little League World Series Rosters (http://www.littleleague.org/series/history/rosters/1981rstr.htm)
[29] Portland Beavers to be sold, likely moved to Escondido, Calif (http://www.oregonlive.com/pbeavers/index.ssf/2010/10/portland_beavers_to_be_sold_mo.html)
[30] MLB.com (http://mlb.mlb.com/news/article.jsp?ymd=20101014&content_id=15635170&vkey=affililiate&c_id=sd)
[31] "Parks, Lakes and Open Space" (http://www.escondido.org/parks-lakes-trails.aspx). City of Escondido. . Retrieved June 8, 2011.
[32] "Grape Day Park" (http://www.escondido.org/grape-day-park.aspx). City of Escondido. . Retrieved June 8, 2011.
[33] "Washington Park" (http://www.escondido.org/washington-park.aspx). City of Escondido. . Retrieved June 8, 2011.
[34] "Ryan Park" (http://www.escondido.org/ryan-park.aspx). City of Escondido. . Retrieved December 2, 2010.
[35] Fikes, Bradley J. (2006-03-05). "Stone Brewing Co. grows in Escondido" (http://www.nctimes.com/business/article_abd07f96-d6aa-5469-b743-2eaa958a6b92.html). North County Times. . Retrieved 2010-05-24.
[36] "Escondido Sister City — Maebaru, Japan" (http://www.escondido.org/escondido-sister-city.aspx). City of Escondido. . Retrieved December 2, 2010.
[37] "City Council" (http://www.escondido.org/city-council.aspx). City of Escondido. . Retrieved December 2, 2010.
[38] Breier, Michelle (December 8, 2010). "Escondido council fills vacancy with former trustee" (http://www.signonsandiego.com/news/2010/dec/08/escondido-council-fills-vacancy-former-trustee/). *The San Diego Union-Tribune*. . Retrieved December 8, 2010.
[39] "City Treasurer" (http://www.escondido.org/city-treasurer.aspx). City of Escondido. . Retrieved December 2, 2010.

[40] "Election Results - County of San Diego - Gubernatorial General Election - Tuesday, November 2, 2010" (http://www.sdcounty.ca.gov/voters/results/transform.htm?paramVal1=city1.xsl). Registrar of Voters - County of San Diego. . Retrieved December 2, 2010.

[41] "Escondido to vote on housing ordinance" (http://www.signonsandiego.com/news/northcounty/20061004-9999-7m4illeg.html). San Diego Union Tribune. .

[42] City delays adoption of parking ordinance (http://www.nctimes.com/articles/2008/09/10/news/inland/escondido/z2a299ee36e4dc838882574c00082a1a9.txt)

[43] Fox, Zach (2008-09-23). "Escondido faces another fiscal obstacle: fewer people" (http://www.nctimes.com/business/article_057ab19c-65c4-5d81-a452-11a8b4ffe8c5.html). North County Times. . Retrieved 2010-05-24.

[44] "Fact Sheet" (http://www.sandag.org/uploads/publicationid/publicationid_52_6470.pdf) (PDF). San Diego Association of Governments. . Retrieved December 2, 2010.

[45] "Will Gerrymandered Districts Stem the Wave of Voter Unrest?" (http://www.clcblog.org/blog_item-85.html). Campaign Legal Center Blog. . Retrieved 2008-02-10.

[46] Escondido Union School District (http://eusd.org)

[47] Escondido Unified High School District (http://www.euhsd.k12.ca.us)

[48] Valley High page at EUHSD (http://www.euhsd.k12.ca.us/districtinfo/intermediate_school_pages/vhs_intermediate.htm)

[49] 2009-10 Accountability Progress Reporting (APR) (http://api.cde.ca.gov/AcntRpt2010/2009Base_Dst.aspx?cYear=&allcds=3768098&cChoice=2009BDst) (Report). California Department of Education. 2009. . Retrieved July 31, 2010.

[50] Notice of Preparation of a Project Environmental Impact Report/Environmental Impact Statement (EIR/EIS) for the California High-Speed Train Project from Los Angeles to San Diego via the Inland Empire, CA (http://cahighspeedrail.ca.gov/images/chsr/20091002134320_NOP_Signed.pdf) (Report). California High-Speed Rail Authority. 2009. . Retrieved July 31, 2010.

[51] "Our Service Territory" (http://www.sdge.com/aboutus/serviceTerritory.shtml). San Diego Gas and Electric. . Retrieved 2010-02-14.

[52] "Our History" (http://www.rinconwater.org/our_history.htm). Rincon del Diablo Municipal Water District. . Retrieved 2010-02-14.

[53] "Palomar Medical Center" (http://www.pph.org/PPHContentPage.aspx?nd=170). Palomar Pomerado Health. . Retrieved 2010-02-14.

[54] Sisson, Paul (June 5, 2010). "Hospital's facade takes shape" (http://www.nctimes.com/news/local/escondido/article_d0aff227-78c0-5568-9f18-953cf4cf9c7b.html). *North County Times*. . Retrieved August 1, 2010.

[55] Angwin, Julia (2009-03-29). "Putting Your Best Faces Forward" (http://online.wsj.com/article/SB123819495915561237.html). The Wall Street Journal. . Retrieved 2010-02-14. "Born Nov. 8, 1970, he grew up in Escondido, a sleepy farm town about a half hour's drive north of San Diego."

[56] Britten, Benjamin (2004). "I. Towards Peter Grimes". In Mitchell, Donald; Reed, Phillip; Cooke, Mervyn. *Letters from a life: the selected letters and diaries of Benjamin Britten.* **Volume three (1946–51)**. Berkeley and Los Angeles: University of California Press. p. 90. ISBN 0-520-24259-9. "Britten and Pears spent much of July and August 1941 in Escondido, staying with the piano duo Ethel Bartlett (1896–1978) and Rae Robertson (1893–1956)."

[57] Eastman, Quinn (2007-05-09). "Escondido a quiet corner of 'D.C. Madam's' life" (http://www.nctimes.com/news/local/escondido/article_4ae2dbd7-5bdd-5199-a4d9-511988044d59.html). North County Times. . Retrieved 2010-02-14. "Deborah Jeane Palfrey, the woman recently accused by federal prosecutors of running a Washington, D.C., prostitution business by phone from California, owns a house in Escondido, but her neighbors said last week that it provided her a quiet refuge, rather than a remote command post."

[58] Lyman, Rick (2000-05-05). "Steve Reeves, 74, Whose 'Hercules' Began a Genre" (http://www.nytimes.com/2000/05/05/us/steve-reeves-74-whose-hercules-began-a-genre.html). *The New York Times*. . Retrieved 2009-07-15. "After he stopped making films in 1969, the couple raised horses at their ranch near Escondido, northeast of San Diego."

[59] Ryan, Travis (January 15, 2009). *Cattle Decapitation* (http://www.thrashermagazine.com/articles/music-interviews/cattle-decapitation/). Interview with Aaron Schmidt. Thrasher Magazine. . Retrieved July 31, 2010.

[60] Sullivan, Mike (July 20, 2007). "Weddle wants to get out of the house" (http://www.nctimes.com/sports/football/professional/nfl/chargers/article_18a5f7f2-752c-51d5-a63a-3f0fc979f0e4.html). *North County Times*. . Retrieved July 31, 2010. "Eric Weddle recently moved into a house in Escondido."

External links

- City of Escondido (http://www.escondido.org/)
- North County Times newspaper (http://www.nctimes.com)
- Escondido Public Library (http://www.escondido.org/library)

Coaster_(San_Diego)

Coaster	
Overview	
Type	Regional rail
Status	Operational
Locale	San Diego County, California, United States
Termini	Oceanside Transit Center Santa Fe Depot
Stations	8
Daily ridership	4,300 (Q1 2010)
Operation	
Opened	February 27, 1995
Owner	North County Transit District
Operator(s)	TransitAmerica Services
Technical	
Line length	41 mi (66 km)
Track gauge	4 ft 8 $\frac{1}{2}$ in (1435 mm)

The **San Diego Coast Express Rail**, or **Coaster** (reporting mark **NCTD**), is a commuter rail service that operates in the Central and Northern coastal regions of San Diego County, California, United States. The service is operated by TransitAmerica Services on contract with North County Transit District (NCTD). The service has eight stops, in San Diego, Solana Beach, Encinitas, Carlsbad, and Oceanside. Service operates primarily during weekday peak periods, but limited weekday midday, Friday evening, Saturday, and seasonal Sunday service also exist. Special evening service is provided when the San Diego Padres play Monday–Thursday evening home games.

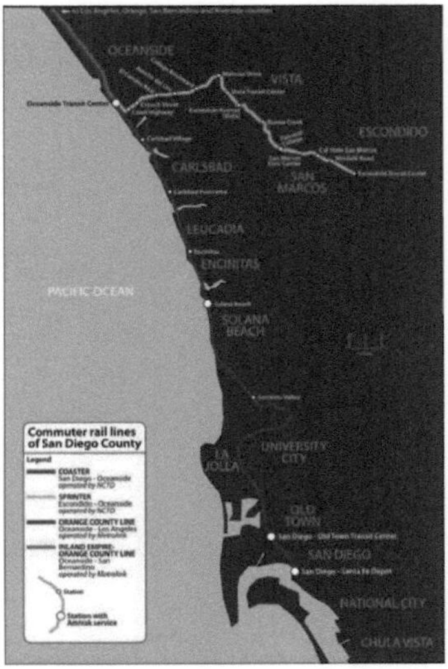

Coaster route map (with other commuter lines included)

History

San Diego Northern Railway (SDNR) purchased the tracks used by Coaster from the Atchison, Topeka and Santa Fe Railway in 1994. NCTD created the subsidiary San Diego Northern Railway Corporation in 1994 and dissolved it in 2002.[1] Revenue Coaster service began February 27, 1995. Money for right-of-way acquisition and construction costs came from TransNet, a 1987 measure that imposed a 0.5% sales tax on San Diego County residents for transportation projects. NCTD originally contracted Amtrak to provide personnel for Coaster trains. On July 1, 2006, TransitAmerica took over the day-to-day operation of the commuter train, based on a five-year, USD 45 million contract with SDNR. TransitAmerica is a subsidiary of Missouri-based Herzog Transit Services.

The Coaster carried 700,000 passengers during its first year of operation. By 2004, it was carrying 1.4 million passengers annually. Approximately 40% of weekday commuters detrain at Sorrento Valley. NCTD owns 62 mi (100 km) of mainline track, as well as the 22 mi (35 km) Escondido Branch that is served by the Sprinter, which began service in March 2008. The Coaster commuter train runs on 41.1 mi (66.1 km) of the mainline.

NCTD maintains two rail yards. One is shared with the San Diego Trolley at 12th & Imperial in Centre City San Diego. It is used to store train-sets during the midday. The other, located north of Oceanside at Stuart Mesa on Camp Pendleton, is shared with Metrolink. This is where the main maintenance facility is located.

Consumption of alcoholic beverages is allowed on board the Coaster for riders who are of the legal drinking age.[2]

Station stops

[3]

Station	Rail Services
Zone 1	
Oceanside	Coaster Pacific Surfliner Metrolink Sprinter
Carlsbad Village	Coaster
Carlsbad Poinsettia	Coaster
Encinitas	Coaster
Solana Beach	Coaster Pacific Surfliner
Zone 2	
Sorrento Valley	Coaster
Zone 3	
Old Town San Diego	Coaster Pacific Surfliner (Limited Weekend Service) San Diego Trolley
Downtown San Diego (Santa Fe Depot)	Coaster Pacific Surfliner San Diego Trolley

Route Description

Stuart Mesa to Oceanside

All Coaster trains start from the Stuart Mesa Coaster Yard, where all of the trains are kept overnight and repaired, just north of the Oceanside Transit Center, the northern terminus of Coaster trains. The train goes over the famed steel bridge that crosses the San Luis Rey River and then goes under the I-5 freeway. The train turns toward the ocean and passes Fallbrook Junction. The train passes a small yard, also called Fallbrook Yard that is not used anymore, except fot the occasional storage of freight cars and Coaster's ballast cars, and then heads for the town of Oceanside. At Oceanside, the train is then in service. The train picks up passengers in Oceanside, its first stop. Then it's on to Carlsbad Village.

Oceanside to Carlsbad Village

The trains leaves Oceanside and passes through a few crossings south of the station. The Sprinter tracks that run next to the Coaster tracks curve inland, as the train passes a wye. It passes CP Longboard, and curves away from the Pacific Ocean. The train then crosses Buena Vista Lagoon, and then shortly after, stops at the Carlsbad Village Stop, the train's 2nd stop.

Carlsbad Village to Poinsettia

After leaving Carlsbad Village, the train rolls through a straight section of track with three more crossings, in the quaint town of Carlsbad. It then crosses Agua Hedionda Lagoon. The train passes by a power plant with a spur rail, that is not used anymore, and crosses Cannon Rd., at high speeds, then through and through CP Farr, southbound trains usually on the siding track, on approach to Carlsbad Poinsettia station stop. The train passes by some houses, and a water treatment plant, and then stops at the Carlsbad Poinsettia Station. It stops at Poinsettia and then rolls on south past CP Ponto.

Poinsettia to Encinitas

After picking up or dropping passengers off at Poinsettia, the Coaster train crosses the Batiquitos Lagoon, with a view of the Pacific Ocean, and then travels under and bridge. The train is now in Leucadia, just north of Encinitas. At high speeds, the train blasts through a grade crossing, and then has to start slowing down, as its next stop, Encinitas is approaching.

Encinitas to Solana Beach

Coaster F40PH-2C #2103 at Encinitas station.

The train then rolls through two grade crossings, and then passes CP Swami. The train then runs next to Highway 101. This view faces directly towards the Pacific Ocean, as the train passes Swami's Beach. The train then passes CP Cardiff and then yet another grade crossing at high speeds as it approaches Cardiff. The train passes Cardiff by the Sea and rolls through a lagoon. The train then rolls through CP Craven and into Solana Beach. The ride into Solana Beach from CP Craven to CP Valley, is mostly below street level in a ditch.

Solana Beach to Sorrento Valley

After leaving Solana Beach, the train passes CP Valley, crosses a lagoon, past the famed Del Mar Race Track, and then rolls through a short siding in Del Mar with CP Crosby on the north end and CP Del Mar on the south end. It passes the abandoned Del Mar Surf Station and then passes another crossing. After travelling on the Del Mar Cliffs, beside the Pacific Ocean, it heads inland. The train rolls through the Soledad Lagoon, and then behind some office buildings in Sorrento Valley. The train then rolls through CP Torrey, stops at Sorrento Valley.

Sorrento Valley to Old Town

After the stop the train passes a grade crossing and rolls past CP Pines. It then travels through the Miramar Hill area that's full of sharp curves. Trains are limited to 25 MPH here. At the top of the hill, the train passes CP Cumbres and the Miramar Wye, and then races down a long siding on the straighter south side of Miramar Hill. It then slows as it reaches CP Elvira. After passing through Rose Canyon, the train starts to pick up speed as it approaches CP Morena. It continues traveling at high speed through CP Tecolote and then slows as it approaches CP Friar and stops at Old Town.

The Coaster and Amtrak station at Solana Beach.

Old Town to Downtown San Diego

After stopping at Old Town, the southbound train passes through several crossings before passing through CP Ash and arriving at the old Santa Fe Depot. Passengers can make transfers to the MTS trolley system (Blue and Orange Lines), MTS Busses, and also the Amtrak Pacific Surfliner, that runs to Los Angeles, Santa Barbara, San Luis Obispo, and points in between. The train is then out of service; it has two choices. On Weekday and most morning trains, the Coaster train can also go to a small storage yard for the day beside the San Diego Trolley yard and the San Diego Imperial Valley yard, next to Petco Park, just south of the Convention Center. Most trains in the non peek hours go back up for the next Northbound Run. Most Coaster Trains stop at Track 1 in Santa Fe Station, and sometimes track 2. Ticket kiosks are located on the platforms, and cannot be bought on board the train.

Ticketing

The cost of Coaster tickets is based upon the number of zones traveled (see map). Tickets must be purchased before boarding, and will occasionally be checked by roving fare inspectors (often between the Solana Beach and Sorrento Valley stations). Monthly passes are available. All tickets and passes include transfer agreements with regular NCTD buses (the "Breeze") and monthly passes include transfer with the Metropolitan Transit System (MTS) buses and Trolleys. As of January 20, 2011, regular one-way fares are as follows:[4]

- Within one zone: $4.00
- Within two zones: $5.00
- Within three zones: $5.50

With proof of eligibility, senior citizens (ages 60 and over), people with disabilities, and Medicare cardholders receive a 50% discount on the above fares.

While tickets are not checked on all trips, penalties for riding the Coaster without a valid ticket include a penalty fare of up to $250. Riders cannot purchase tickets on board the train. Monthly passes are available for between $120.00 to $165.00. With proof of eligibility, youth (ages 6–18) monthly passes are $82.50, while senior, disabled and Medicare monthly passes are $41.25.

In January, 2011, Coaster introduced lower fares to encourage more customers to take the train.

Schedule

Monday to Friday (Weekdays)

Coaster runs thirteen daily southbound trains (Oceanside to San Diego) and thirteen northbound (San Diego to Oceanside) weekdays, Monday to Friday. Three Coaster trains, the morning trains, then are stored in the MTS Trolley yard, just south of the main San Diego Depot. In the afternoons, the trains come back to the station, load, and then continue on its normal route. On Friday and Saturday nights, Coaster operated an additional four trains to accommodate Friday night crowds during the summer months. Coaster stops this service when the Fall timetable switches.

Saturday and Sundays (Weekends)

Saturdays have limited service, with six southbound and six northbound trains in the summertime. In time Fall and the Winter, the weekend Saturday trains go down to 4 trips Southbound, and also 4 trips Northbound. On April 4, 2011, Coaster introduced a limited 4-roundtrip Sunday/holiday service during the summer season. This then was extended to be year round. Special service to Padres games is available in the summertime. Special Service for Special Events, such as the Carlsbad Street Fair, also have more trains running on the weekend.

Compass Card

In September 2008, SANDAG introduced a new contactless "Compass Card", made possible by Cubic Transportation Systems, Inc.. The "Compass Card" allows passengers from MTS and NCTD to store regional transit passes and cash value on a rewritable RFID card. Customers can purchase passes and add cash value on the Internet or at any ticket vending machine. When a customer boards a bus they simply tap their Compass Card on the "Validator", usually located near or on the farebox. The LED display then lights up with lights resembling that of a stoplight, and the LCD display shows text regarding the passenger's fare account. When boarding rail vehicles the Validators are located on the platform, and the same process is performed to board the train. This program is still in the pilot testing phase, and SANDAG hopes to have all bus passengers using the Compass Card by the end of 2009.

Fleet

Coaster F40PHM-2C locomotive #2104 with Amtrak Pacific Surfliner #572 behind it at Solana Beach Station, California

Builder	Type	Purchased	Numbers
Morrison-Knudsen	F40PHM-2C	1994	2101–2105
GM-EMD	F59PHI	2001	3001–3002
Bombardier	Bi-Level Coach	see below	see below

Trailer numbers are 2201–2208 (1994), 2401–2406 (1997), 2501–2504 (2003)

Cabcar numbers are 2301–2308 (1994), 2309–2310 (2003)

Most trains run with 5 coaches. However, as of July 2008, a 6th coach has been added to half of the weekday trains due to an 8% jump in ridership from the previous year.

Bike rider information

2401–2406 series coaches have space for 6 bikes (4 on north end, 2 on south end).
2501–2504 series coaches have space for 6 bikes (all in the south end of the coach).

Connecting rail and bus transit services

The Coaster connects with Amtrak at Oceanside, Solana Beach, and San Diego. It also connects with Metrolink at Oceanside. The Coaster connects to the San Diego Trolley and buses at Old Town and Santa Fe Depot in San Diego, including to the Lindbergh Field. The Sprinter line connects Oceanside, California to Escondido, California. Riders can use buses from the Sorrento Valley station to access employment centers in Sorrento Valley, Sorrento Mesa, Torrey Pines, and University City.[5] The VA Hospital, the University of California, San Diego, and some area employers also provide shuttle service to and from the Sorrento Valley station at peak times. NCTD provides "Breeze" bus service connecting with the Coaster at all North County station stops.

Future prospects

The Coaster remains stable for the foreseeable future. San Diego County voters extended the TransNet sales tax through 2038, which includes some funds for track upgrades. Plans to double track more of the NCTD-owned mainline are also being considered. There are also plans of extending Coaster service to Camp Pendleton and the Convention Center,[6] as well as interest in rebuilding the station at Del Mar to serve the racetrack and fairgrounds.[7]

Yards

The Coaster has 2 storage yards. The main storage yard is in Stuart Mesa, just north of the Oceanside station stop. This is where all of the cars are stored for the night, and where the trains get serviced. A small yard beside the San Diego Trolleys' yard is for staging of trains for the day, but only is used during the weekdays. Three trains are parked there during the day.

See also

- Transportation in San Diego County

External links

- Official NCTD site [8]
- 511sd.com—New site for sdcommute.com [9]
- San Diego Metropolitan Transit System [10]

References

[1] "North County Transit District-We Move People" (http://www.gonctd.com/pdf/srtp/ch2.pdf). North County Transit District. .

[2] Sisson, Paul (2 Aug 2006), "Brawl leads some to question Coaster drinking policy" (http://www.nctimes.com/articles/2006/08/03/news/coastal/8206190449.txt), *North County Times*, , retrieved 4 Mar 2009

[3] Fare Reduction - Jan. 20, 2011 (http://gonctd.com/pdf_alerts10/COASTER drop.pdf)

[4] http://www.gonctd.com/coaster_fares.htm

[5] Coaster Connecting Services (http://www.gonctd.com/connecting_services)

[6] "Marines, NCTD eye Camp Pendleton Coaster stop" (http://www.signonsandiego.com/news/2011/nov/11/marines-nctd-eye-camp-pendleton-coaster-stop/). *The San Diego Union-Tribune*. November 11, 2011. . Retrieved November 15, 2011.

[7] St John, Alison (2008-03-14). "SANDAG Board to Explore Viability of Del Mar Track Train Station" (http://web.archive.org/web/20080716022145/http://www.kpbs.org/news/local;id=11166). KPBS. Archived from the original (http://www.kpbs.org/news/local;id=11166) on 2008-07-16. . Retrieved 2008-08-03.

[8] http://www.gonctd.com/coaster_intro.htm

[9] http://transit.511sd.com/Default.aspx

[10] http://www.sdmts.com/

Palomar_College

Palomar College	
Motto	Learning for Success
Established	1946
Type	Public
President	Robert P. Deegan
Academic staff	1,100 as of Fall 2003
Undergraduates	30,000
Location	Main: San Marcos, CA, USA
Campus	Main: Urban; 200 acres (81 ha)
Athletics	24 teams
Colors	Scarlet and Silver
Mascot	Comets
Website	www.palomar.edu [1]

Palomar College is a community college with one campus and six education sites in San Diego County, California. The main campus is located in San Marcos, while the six education sites are located elsewhere throughout north San Diego County. The largest of these is the education center located in Escondido. Other education sites are located at Camp Pendleton, Mt. Carmel High School, Fallbrook High School, the Pauma Indian Reservation, and at Ramona High School.

The Palomar Community College District's facilities improvement measure, Proposition M, was passed by 57% of voters in the November 8, 2006 General Election. As a result, the $694 million provided by the measure, as well as over $200 million in matching funds from the State and an additional $37 million from Proposition 1D, will provide for the implementation of the college's Master Plan 2022 [2].

In 2009, the college developed an "Integrated Planning, Evaluation, and Resource Allocation Decision-Making Model (IPM). This IPM aligns the college's long-range Master Plan, its mid-range Strategic Plan, and its short-range Program Review and Planning processes while also incorporating the new Resource Allocation Model (RAM).[3]

Academic programs

Palomar College offers more than 250 associate's degrees and certificate programs, and also offers programs for students wishing to transfer to many different four-year universities, including institutions in the University of California and California State University systems.

- Academic Divisions:
  - Arts, Media, Business and Computer Systems
  - Career, Technical and Extended Education
  - Languages and Literature
  - Mathematics and The Natural and Health Sciences
  - Social and Behavioral Sciences.

Media, Newspapers and Magazines

- Palomar College Newspaper - The Telescope [4]
- Palomar College Radio - KKSM [5]
- Palomar College Television - PCTV [6] (shares airtime with ITV Cable 16)

Associated Student Government

The Associated Student Government is a student run organization at Palomar that strives to create a better campus for its students. Members of the ASG serve on campus-wide shared-governance committees and hiring committees, lobby State and Federal representatives on student issues, attend leadership conferences, and are responsible for Comet Week, Springfest, and some campus-wide activities. Members of the ASG have opportunities to attend conferences, special on-campus parking, and serve on campus wide committees as the "voice of the students."

Fall 2011/Spring 2012

- President: Evelyn Lucero
- Vice President: Cody Dean

Fall 2010/Spring 2011 Election results:

- President: Channing Shattuck
- Vice President: Evelyn Lucero

Fall 2008/Spring 2009 Election results:

- President: Robert Frederick
- Vice President: Claudia Duran
- ICC Chairperson: Jessica Baker

Buildings

List of Buildings

Name	Full Name	Constructed	Remodeled	Demolished	Renamed
A	Administration	1960	1966, 1987, 1992		
AA/ST	Administration Annex	1974	1976, 1978, 1988		
AF	Adjunct Facilities				
B	Business	1960	1978, 1989	2011	
BE	Behavioral Science	1978	2011		
BES	Behavioral Science A	1978			
C	Art Complex	1966	1978, 1994		
CCC	Child Care Center	1975	N/A	N/A	1993
CDA	Child Development #1	1983	1978, 1994		
CDB	Child Development #2	1975	1993		
CDC	Child Development #3	1972			
CES	California English School	1992			
CH	Chemistry	1965	1988	2008	
CP	Campus Police	1993	2004		
CS					
CT	Racquetball Courts & Weight Training Center	1976	2002		
D	Music	1966	1979, 1993		
DSP&S	Disabled Resource Center	1974	1986		
E	Math	1960	1993		
EL					
ES	Earth Science	1965	1978	2008	
ET	Educational Theatre	1979	N/A	N/A	1983
ETV	Educational Television				
F	Foreign Language & Photography	1960	1964, 1989		
FASH	Fashion Design	1992			
FCS	Family & Consumer Sciences	1993			
G (Dome)	Gymnasium	1958	2000		
GJ	Graphics Communications	1978	1988		
GMS					
GN					
HBT	Howard Brubeck Theatre	1979	2000, 2011-13		
HC	Health Center	1978	2011		
HH	Health Services	1974	N/A	N/A	1986
HS	Health Sciences	2009-10			
HUM	Humanities	2011-13			
IRIS					
IS					

IT	Industrial Technology	1956	1965, 1966		
ITC	Industrial Technology Center	2010-11			
J	Maintenance Complex	1965			
L	Library	1956	N/A	N/A	1985
LL	Library and Learning Resource Center	1983	1999		
LS	Life Science	1965	1978	2008	
M	Men's Locker Room	1955	1960, 1964		
MB	Modular Buildings	2008	N/A	2011 (15/16)	
MD	Multidisciplinary Instructional	2009-11			
N	Auto Shop	1967			
NA	Nursing Classroom A	1978			
NB	Nursing Classroom B	1978			
NO	Nursing Offices	1978	2011	N/A	2011
NS	Natural Sciences	2005-07			
O	Women's Locker Room	1965			
P	Humanities/Social Sciences/ETV	1965	1972, 1976, 1993	2011 (north)	
PA					
PAO	Public Affairs Office				
Q	Electronics and Telecommunications	1965			
R	Student Union/Café/ Bookstore	1958	N/A	N/A	??
RC	Reading Center	1978	1983, 1992		
RF	Reading and Food Services	1992			
RS	Receiving and Shipping	1974	1985, 1992		
S	Science I	1956	N/A	2008	
SC	Speech and Communications	1978			
SHS	Student Health Services	1986	N/A	2012	
SSC	Student Services Center	1956	1964, 1985		
STB					
STM					
STX					
SU	Student Union/Café/Bookstore	1958	1964, 1975, 1978, 1980, 2002-03, 2011	2002 (Old)	
SW	Swimming Pool Complex	1971	1976		
T	Trades & Industry	1979	2012		
TCA	EOP&S	1978			
TCB	TRIO & Telescope	1978			
TLC	Teaching Learning Center	2011-13			
TS					
TSA					
TT					

U	Communication Offices	1972			
W	Behavioral Sciences Offices	1972	N/A	2011	
WFC	The Wellness Fitness Center	1994			

Clubs

There are many student run clubs at Palomar College, including:

- Anime Club [7]
- Archaeology Club [8]
- The Fellowship of Christian Athletes [9]
- Palomar LGBTQ & Allies [10]
- Geoscience Connection [11]
- Palomar College Democrats [12]
- Phi Theta Kappa [13]
- Photography Club [14]
- Pre-Med Club [15]
- Speech & Debate Club [16]
- Environmental Student Organization [17]

Athletics [18]

- Men's and women's
  - Basketball
  - Volleyball
  - Tennis
  - Soccer
  - Swimming
  - Water polo
  - Cross country
- Other Sports
  - Football [19]
  - Baseball
  - Softball
  - Golf
  - Wrestling
  - Track and Field.

Notable alumni

- Tom Dempsey - Former NFL kicker best known for his record 63 yard field goal with the New Orleans Saints.
- Candi Kubeck - Captain of ValuJet Flight 592, which crashed into the Florida Everglades on May 11, 1996.
- Matt Chico - MLB pitcher currently with the Washington Nationals.
- Ken Block - Rally Car Driver, and Founder of DC Shoes.
- Brett Salisbury - College football quarterback and author of the Transform Diet.
- Tom Luginbill - College football quarterback and college football analyst for ESPN.

References

[1] http://www.palomar.edu/
[2] http://www.palomar.edu/masterplan/
[3] Palomar Community College District Accreditation Follow-Up Report. March 2010. (http://www.palomar.edu/accreditation/)
[4] http://www.the-telescope.com
[5] http://www.palomar.edu/kksm/
[6] http://www.palomar.edu/pctv/
[7] http://clubs.palomar.edu/anime
[8] http://clubs.palomar.edu/archaeology
[9] http://www.socalfca.org/
[10] http://www.palomar.edu/clubs/LGBTQ/
[11] http://clubs.palomar.edu/geoscienceconnection
[12] http://palomarcollegedemocrats.org/
[13] http://clubs.palomar.edu/phithetakappa
[14] http://clubs.palomar.edu/photography
[15] http://clinic4mexico.googlepages.com/
[16] http://www.palomar.edu/speech_and_asl/
[17] http://pc-eso.wikidot.com/
[18] http://www.palomar.edu/athletics/
[19] http://www.palomar.edu/athletics/football/

External links

- Palomar College Website (http://www.palomar.edu/)
- Palomar Community College Library (http://www.palomar.edu/library/)
- Palomar Community College Forums (http://palomarforums.co.nr)
- Palomar Community College Newspaper - The Telescope (http://www.the-telescope.com)
- Palomar Community College Radio - KKSM (http://www.palomar.edu/kksm/)
- Palomar Community College Student Government - Associated Student Government (http://www.palomar.edu/asg/)
- http://www.palomar.edu/masterplan/PAL%20MP%20Existing.pdf Existing District Facilities - List of Buildings Table

California_State_University,_San_Marcos

California State University, San Marcos	
Established	1989
Type	Public
Endowment	US$13.0 million (as of June 30, 2009)[1]
President	Karen S. Haynes[2]
Provost	Emily F. Cutrer [3]
Academic staff	246[4]
Students	10,276 (Fall 2011)[5]
Location	San Marcos, California, United States
Campus	Suburban, 340 acres (140 ha)
Colors	Pantone Blue, Black, White
Athletics	NAIA Region II
Nickname	Cougars
Mascot	Tukwut [6]
Affiliations	AASCU California State University
Website	CSU San Marcos [7]

CELEBRATING 20 YEARS

California State University San Marcos (also **CSUSM** or **Cal State San Marcos**) is a public, coeducational university and one of the 23 general campuses of the California State University system.[8] located in San Marcos, California, a suburban town in north San Diego County. It was founded in 1989 as the 20th CSU campus and was the first after nearly 30 years. The first class was admitted in 1990.

CSU San Marcos offers 44 undergraduate programs,[9] 10 graduate programs and a Doctorate in Education[10] in three colleges and one nursing school.[11]

History

Efforts by community and political leaders to attract a state university to North County date back to 1968. In 1978, state legislator William A. Craven (1921–1999) won state funding for a North County satellite campus of San Diego State University. In 1989, Gov. George Deukmejian signed another Craven bill which established a CSU campus in San Marcos.

The state purchased land in San Marcos, including the former Prohoroff Poultry Farms chicken ranch. The hillside site lies approximately 8 miles (13 km) due east of the Pacific Ocean and 35 miles (56 km) due north of downtown San Diego. Today the campus comprises 340 acres (140 ha).

Bill Stacy was appointed president in June 1989. During the 1989-1990 academic year, Stacy hired 12 "Founding Faculty," who played an important role in the university's early years and today are memorialized in Founders Plaza.

CSUSM admitted upper division students and held classes from September 1990 through August 1992 in rented facilities in a San Marcos business park, alongside the North County campus of San Diego State University. Groundbreaking for the permanent campus occurred on February 23, 1990. Classes began at the current campus in August 1992. Craven Hall was the third instructional building opened in December 1992, and the largest building up to that time at 155000 square feet (14400 m^2).

Since 1992, the campus has been growing steadily. Major additions over the subsequent decade include the Foundation Classroom Buildings (December 1996), University Hall (1998), a second Science building (August 2002) and the Arts building (August 2002). The largest building to date is Kellogg Library, opened January 2004 with nearly 200000 square feet (19000 m^2). The most recent addition is Markstein Hall (January 2006), home of the College of Business Administration.

For the 2011-2012 academic year, tuition and fees rose to $6,596, a 31% increase attributed to the state's budget crisis; it is the largest such percentage increase in the United States.[12]

Academics

Originally, the university admitted only upper division students; enrollment in 1990-1991 was 448. The first freshman students entered in 1995, when enrollment totaled 3,642.

The university's enrollment (as of Fall 2009) was approximately 9,700 students; about two-thirds of the students were from San Diego County. The campus master plan calls for an eventual enrollment of 25,000.

The university has three colleges: Arts & Sciences, Business Administration, and Education as well as a School of Nursing in addition to a robust Extended Learning program.

- College of Arts and Science (COAS)
- College of Business Administration (CoBA)
- College of Education (COE)

In less than two decades, the campus has distinguished itself. CSU San Marcos' programs in teacher education are renowned for their school and community collaboration. The College of Business Administration (COBA) receives national attention for the “Senior Experience” [13] program, which takes teams of students off campus for projects with companies and organizations. Programs in the sciences and in visual and performing arts are regionally acclaimed for academic rigor, innovation, and career preparation.[14]

The university is fully accredited by the Western Association of Schools and Colleges (WASC) for both Bachelors and Masters degree programs, receiving initial accreditation in 1993. The university is undergoing its seven year review, having just completed the Capacity and Preparatory Review visit in March 2007, and is expected to finalize the process in spring 2009 following the Educational Effectiveness Review. The school reduced the number of units needed to graduate. Some have criticized this policy change stating that students will not receive a liberal and broad education.

Campus culture

2011 Demographics of student body

	Undergraduate
African American	2.8%
Asian American	10.8%
White American	49.5%
Hispanic American	21.5%
Native American	0.9%
international	1.8%
Ethnicity unreported/unknown	12.7%

A large majority of the student population comes from the San Diego and Riverside County areas. Approximately 600 students live in on-campus residence halls. Co-curricular programs and activities are sponsored by Associated Students, Incorporated, Student Life and Leadership, the Clarke Field House/University Student Union, and the University Village Apartments. There are more than 100 student organizations. The school attracts many students from the surrounding community colleges including Palomar College and MiraCosta College, as well as from others located in North San Diego County, Southern Orange County, and Southern Riverside County.

The school paper is called "The Pride [15]."

Cal State San Marcos regularly competes in RecycleMania, a college and university recycling program and competition.[16] In 2010, for the sixth straight year, Cal State San Marcos won first place in the nationwide RecycleMania contest.[17]

According to a local newspaper[18] and the 2007-2008 Clery Report on Crime Awareness and Campus Security,[19] CSU San Marcos is a safe campus with a relatively low crime rate. The University's police department employs a variety of crime prevention and safety education programs to keep the campus safe.[19] In 2008, seven gang members were arrested at an off-campus, privately-operated apartment complex located next to the University. Originally, there were concerns that this gang was selling drugs to students. However, as a result of the efforts of the local sheriff's department and the apartment's new security team, the gang appears to have vacated the apartment complex.[20]

Athletics

The original mascot of the campus was established as the "Tukwut," the name for the California mountain lion in the language of the Luiseño Native American tribe. However, a 1999 student referendum selected the more generic "Cougars" as the name for all CSUSM sports teams.

The school's athletics department has ten teams (women's and men's in each of the following sports): golf, cross country, track & field, soccer, as well as women's softball, and men's baseball. The official colors of the Cougars are bright/royal blue and white. In the early years, burgundy was used sparingly as an accent. Cougar athletic teams compete as independents in Region II of the NAIA. The school recently hired their first coaches in baseball and soccer; in baseball, Dennis Pugh, who coached at a local high school for decades, and soccer coach Ron Pulvers. In 2007, the softball program's second year, the school hired former UCLA All-American Kelly Warren who was previously the Associate Head Coach at nearby San Diego State University. Steve Scott (aka "The Miler") is the Cougars' Cross Country and Track & Field coach. In Fall 2009, the Women's Cross Country team won their first NAIA National Championship. They went on to defend their title the next year in Fall of 2010. The university also has a surf team that competes in the National Scholastic Surfing Association (NSSA). The surf team won the NSSA collegiate national championship in 2009. For the academic year of 2011, women's volleyball, and men's and women's basketball, have all been added to the campus.[21]

President and faculty

The university has had four presidents:

- Bill Stacy (1989–1997), who left to become chancellor of the Chattanooga campus of the University of Tennessee
- Alexander Gonzalez (1997–2003), previously the provost of the California State University, Fresno who left CSUSM to become president of the larger California State University, Sacramento
- Roy McTarnaghan (2003–2004), interim president
- Karen S. Haynes, (2004-), former president of University of Houston–Victoria, who joined CSUSM in February 2004

The university opened in 1990 with the 24 faculty. Today it employs 200 tenure-track and tenured faculty out of a total of 980 employees.

References

[1] "U.S. and Canadian Institutions Listed by Fiscal Year 2009 Endowment Market Value and Percentage Change in Endowment Market Value from FY 2008 to FY 2009" (http://www.nacubo.org/Documents/research/2009_NCSE_Public_Tables_Endowment_Market_Values.pdf) (PDF). *2009 NACUBO-Commonfund Study of Endowments*. National Association of College and University Business Officers. . Retrieved February 2, 2010.

[2] "Welcome to Cal State San Marcos!" (http://www.csusm.edu/president/). *CSUSM Office of the President*. CSUSM.edu. . Retrieved 2011-10-28.

[3] http://www.csusm.edu/aa/provostoffice.htm

[4] Headcount of Full-Time Faculty by Rank and Campus, The California State University web site (http://www.calstate.edu/hr/employee-profile/faculty/campus.shtml)

[5] "Enrollment at CSUSM Reaches Record High" (http://www.csusm.edu/news/topstories/current/tsFall2011Enrollment.html). CSUSM.edu. 2011. . Retrieved 2011-10-28.

[6] http://www.csusm.edu/communications/topstory/tsTukwut.html

[7] http://www.csusm.edu/

[8] "Campus Homepages" (http://www.calstate.edu/search_find/campus.shtml). *California State University*. . Retrieved 2008-11-07.

[9] "Academic Offerings" (http://www.csusm.edu/academics/majorsminors.html). CSUSM.edu. . Retrieved 2011-10-28.

[10] http://www2.csusm.edu/rgsip/graduate_studies/index.htm

[11] "CSU San Marcos Colleges & Programs" (http://www.csusm.edu/academics/colleges.html). CSUSM.edu. . Retrieved 2011-10-28.

[12] "5 biggest state tuition hikes" (http://money.cnn.com/galleries/2011/pf/college/1110/gallery.tuition_increase/). *CNNMoney*. October 26, 2011. . Retrieved 2011-10-28.

[13] http://www.csusm.edu/seniorexperience

[14] "Why choose CSU San Marcos?" (http://www.csusm.edu/admissions/whycsusm/index.html). CSUSM.edu. . Retrieved 2011-10-28.

[15] http://www.thecsusmpride.com/

[16] "California State University-San Marcos" (http://www.recyclemaniacs.org/university_detail08.asp?ID=3813). *Participating Schools*. RecycleMania. . Retrieved 2011-10-28.

[17] "Final Results—RecycleMania 2010" (http://www.recyclemaniacs.org/results-2010.htm). RecycleMania. 2010. . Retrieved 2011-10-28.

[18] http://www.nctimes.com/articles/2008/11/02/news/sandiego/z8427a22d53505855882574f200648a6d.txt

[19] http://www.csusm.edu/police/Clery_Report_08Final.pdf

[20] http://www.northcountytimes.com/articles/2008/11/18/news/inland/san_marcos/zf892b43cad32ef7c882575010073211d.txt

[21] "CSUSM Surf Team Claims National Championship" (http://www.csusm.edu/news/articles/nr_surfteamchampions.html) (Press release). California State University, San Marcos. September 15, 2009. . Retrieved April 7, 2011.

External links

- Official website (http://www.csusm.edu/)

North_County,_San_Diego

North County

— Region of the San Diego Metro Area —

Oceanside Pier in North County

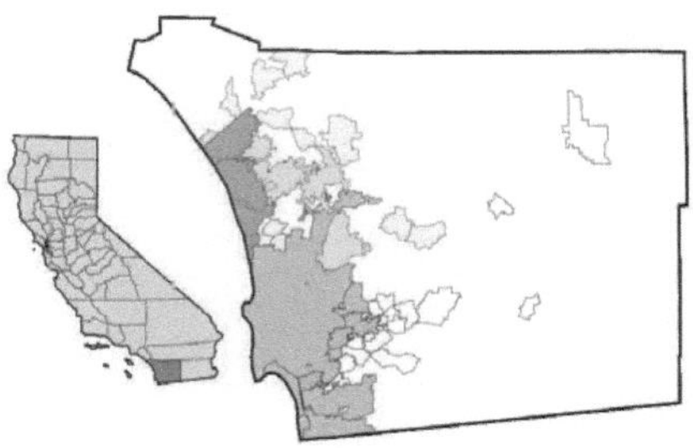

North County Coastal and Inland cities

County	San Diego County
Population	821496

North County is a region in the northern area of San Diego County, California. It is the second most populous region in the county after San Diego, with an estimated population of 826,985. North County is well known for its affluence, especially in Cardiff-by-the-Sea, Carlsbad, Del Mar, Rancho Santa Fe, and Solana Beach where house prices range on average above one millions dollars. [1]

History

The name dates to at least the 1970s, when many of the communities in the area were yet to become incorporated cities and local community decisions were made 40 miles away at the county seat. The North County section of San Diego County has historically been the most expensive region of San Diego, with such affluent neighborhoods as Cardiff-by-the-Sea, Carlsbad, Carmel Valley, Del Mar, Encinitas, Olivenhain, Rancho Santa Fe, and Solana Beach.[2]

In modern times North County continues to grow as a highly influential region of Greater San Diego. The top twenty-five employers in San Diego County are closer to the North County city of Carlsbad than San Diego proper.[3]

Geography

Both coastal and inland North County maintain two types of topography. In North County Coastal the land is generally flat with low rolling hills. The beaches are sandy with occasional tidepools and rocky reefs popping out of the surf. In some cases the coast is dominated by bluff type geography, where the land meeting the ocean sharply drops into the sea with a short beach. In some cases, such as in Encinitas, a whole city can be bisected by a coastal foothill ridge. Mountains soon become imminent as one travels further inland, displaying the rocky peaks of North County Inland. Such peaks included Twin Peaks in Poway, Mount Woodson in Ramona, and Iron Mountain in unincorporated territory between the two cities.

Rivers and creeks flowing west from the mountains farther inland predominantly end up draining into the regions four main lagoons. Throughout their course the way they form many lakes and reservoirs supporting and array of native species.

Definitions vary, but almost always include the communities and cities along Interstate 5 north of Carmel Valley Road and Interstate 15 north of Lake Hodges. For both geographic and political reasons, opinions differ as to whether to include the northern communities within the city limits of San Diego (such as La Jolla, Rancho Bernardo, or Scripps Ranch), or close in communities and cities such as Solana Beach, Cardiff-by-the-Sea, Aviara, Del Mar, Carmel Valley, Rancho Santa Fe, etc.

Regardless, North County is commonly divided into coastal and inland regions. The North County Times, which serves the entire region, shows North County Coastal cities being Oceanside, Carlsbad, Vista, Encinitas, Del Mar, and Solana Beach and North County Inland cities being Escondido, Fallbrook, San Marcos, Poway, Valley Center, Ramona, and Rancho Bernardo.[4]

Ecology

The region has strong ties to its coast. Its notable efforts are popular for preserving many marine environments including lagoons and tidal wetlands, many being last few on the South Coast.[5] Unlike developments in many Orange County coastal cities, the lagoons and large areas of coast have not been so heavily developed. Major lagoons and inlets lining the coast from north to south include: Oceanside Harbor, Buena Vista Lagoon, Agua Hedionda Lagoon, Batiquitos Lagoon and San Elijo Lagoon.

California Least Tern

Flora and fauna

These lagoons provide valuable wetland habitat for many bird, reptile, fish, and plant species. The waters off the coast are also very rich in species diversity; supporting large kelp forests and rocky reefs.

Fish species included the tidewater goby, topsmelt, striped mullet, surfperch and Pacific staghorn sculpin. Leopard sharks forage near the lagoons and pups frequent the shallow rocky reefs off the coast.

Bird species included the Great Blue Heron, Snowy Plover, Clapper Rail and Least tern. The lagoons support various species of shorebirds, wading birds, waterfowl, raptors and diving birds.[5] The number of bird species in the San Dieguito Wetlands have tripled due to restoration projects by Del Mar.[6]

Culture

North County Coastal city of Del Mar

North County contains forty golf courses, including Torrey Pines, which hosted the 2008 U.S. Open.[7] North County is home to Southern California's only five star and five diamond restaurants: Addison at The Grand Del Mar, and El Bizcocho at Rancho Bernardo Inn.[7] The Del Mar Fair is home to one of the most famous racetracks in the world and is the site of the annual San Diego County Fair. Shopping malls include the North County Fair in Escondido and Plaza Camino Real in Oceanside. Famous beaches include Moonlight Beach in Encinitas and the Oceanside Pier.

A notable fictional character from North County is Dave Rickards' "Aunt Edna," frequently featured on the popular *Dave, Shelly, and Chainsaw* radio program, which airs in the San Diego area.

Demographics

North County is considered wealthier and more socially conservative than other parts of the metropolitan area such as South Bay. It is the second largest region of the San Diego metropolitan area. There are 325,995 people residing in North County San Diego. 49.1% are males versus 51.9% female. White collar jobs beat blue collar jobs 3:1. More people have B.A. Degrees then AS/AA. degrees 2:1. More people are married then single 2:1.Demographs [8]

Cities

Incorporated cities

North County Coastal city of Oceanside

Populations listed are from the 2010 United States' Census

- Oceanside - 167,086
- Escondido - 143,911
- Carlsbad - 105,328
- Vista - 93,834
- San Marcos - 83,781
- Encinitas - 59,518
- Poway - 47,811
- Solana Beach - 12,867
- Del Mar - 4,161

Unincorporated CDPs

Populations listed are from the 2010 U.S. Census

North County Inland city of Escondido

- Fallbrook - 30,534
- Ramona - 20,292
- Camp Pendleton South - 10,616
- San Diego Country Estates - 10,109
- Valley Center - 9,277
- Camp Pendleton North - 5,200
- Lake San Marcos - 4,437
- Hidden Meadows - 3,485
- Bonsall - 3,982
- Fairbanks Ranch - 3,148
- Rancho Santa Fe - 3,117
- Rainbow - 1,832

Education

Primary and secondary

Colleges and universities

North County is home to several colleges, both traditional four year universities and community colleges. These included California State University, San Marcos (CSUSM) offering undergraduate, graduate, and doctorate programs;[9] and the community colleges Palomar College and MiraCosta College.

Media

Newspaper

The *North County Times* is the main newspaper in the region.

Radio

Landmarks

Important landmarks in North County include Del Mar Racetrack, Lake San Marcos, Oceanside Pier, Stone Steps, Twin Peaks

Del Mar Race Track

Lake San Marcos

Oceanside Pier

Twin Peaks

Politics

Most of North County is located within County Supervisorial District 5, and is represented on the county's Board of Supervisors by Bill Horn. The remainder of North County is in Districts 2 and 3, represented by Dianne Jacob and Pam Slater-Price, respectively.

See also

- East County
- South Bay

References

[1] "San Diego Real Estate Market Reports" (http://www.sandiegorealestatehq.com/north-county-real-estate-report.php). Highland Realty. . Retrieved March 10, 2011.
[2] NPR (http://www.npr.org/templates/story/story.php?storyId=5031130)
[3] California Pacific Airlines. "San Diego North Means Business" (http://www.flycpair.com/landingpage/business.htm). . Retrieved 2011-03-03.
[4] North County Times: Coastal and Inland cities (http://www.nctimes.com/)
[5] "Batiquitos Lagoon" (http://www.batiquitosfoundation.org/newsite/index.php). Batiquitos Lagoon Foundation. . Retrieved 2011-03-03.
[6] "DEL MAR:Restoring the wetlands" (http://www.nctimes.com/news/local/del-mar/article_64da19e1-65cc-5a4f-90ad-dd85422905e2.html). North County Times. . Retrieved March 10, 2011.
[7] San Diego's North County (http://www.sandiego.org/article/Visitors/119)
[8] http://homes.point2.com/Neighborhood/US/California/San-Diego-County/San-Diego/North-San-Diego-County-Demographics.aspx
[9] "CSUSM: CSU San Marcos Academics" (http://www.csusm.edu/academics/). California State University San Marcos. . Retrieved June 12, 2011.

External links

- San Diego North Convention & Visitors Bureau (http://www.sandiegonorth.com/)
- City of Oceanside (http://www.ci.oceanside.ca.us/)
- City of Escondido (http://www.escondido.org/)
- City of Carlsbad (http://www.carlsbadca.gov/Pages/default.aspx)
- City of Vista (http://www.cityofvista.com/)
- City of San Marcos (http://www.ci.san-marcos.ca.us/)
- City of Encinitas (http://www.cityofencinitas.org/)
- City of Poway (http://www.poway.org/index.aspx?page=1)
- City of Solana Beach (http://www.ci.solana-beach.ca.us/csite/cms/home.htm)
- City of Del Mar (http://www.delmar.ca.us/default.aspx)

North_County_Transit_District

North County Transit District (NCTD)	
NCTD	
Info	
Locale	North San Diego County, California
Transit type	Multi-Modal regional transit network consisting of: Buses Light Rail Commuter rail & Paratransit
Operation	
Began operation	September 30, 1975
Operator(s)	Breeze & Lift: NCTD Sprinter: Veolia Transportation Coaster: TransitAmerica Services
Reporting marks	SDNR
Technical	
Track gauge	4 ft 8½ in (1435 mm) (standard gauge)

The **North County Transit District** (**NCTD**) (reporting mark **SDNR**) is the agency responsible for public transportation in North San Diego County, California, United States. The agency operates the "Breeze" bus service, the Sprinter light rail service between Oceanside and Escondido, and manages the San Diego Coaster, a commuter rail service between Oceanside and Downtown San Diego, in conjunction with Amtrak.

NCTD owns 62 mi (100 km) of mainline track, as well as the 22 mi (35 km) Escondido Branch that is served by the Sprinter beginning in 2008. The Coaster commuter train runs on 41.1 mi (66.1 km) of the mainline.

NCTD maintains two rail yards. One is shared with the San Diego Trolley at 12th & Imperial in Centre City San Diego. It is used to store trainsets during the midday. The other, located north of Oceanside at Stuart Mesa on Camp Pendleton, is shared with Metrolink. This is where the main maintenance facility is located.

History of the San Diego Northern Railway and NCTD

San Diego Northern Railway (SDNR) purchased the tracks used by Coaster from the Atchison, Topeka and Santa Fe Railway in 1994. NCTD created the subsidiary San Diego Northern Railway Corporation in 1994 and dissolved it in 2002.[1] Total investments through NCTD total $750 million.

In November 2009, NCTD approved outsourcing all bus and rail operations effective July 1, 2010 to First Transit. The agency would save $70 million over seven years with the move.[2] Fleet and facility operators remained NCTD employees until their contracts expired June 30, 2011 [3]

NCTD still uses the SDNR reporting marks for its rail equipment. With new rail operations management, between January 2010 and June 2011 ridership increased 29%. This is the single greatest increase in ridership for all US rail-lines in the year 2010.

Services

- **Breeze**- Fixed-route local and express bus services.
- **Breeze Rapid** - Bus rapid transit line in Escondido.
- **Coaster**- Commuter rail service from Oceanside to Downtown San Diego.
- **Lift**- Paratransit service for those with disabilities that are unable to use the accessible fixed-route system.
- **Sprinter**- Light rail from Oceanside to Escondido, California.

Breeze Lines

As of January 2011, there are thirty bus routes run by NCTD.

- **101** - Oceanside Transit Center to UTC via former Highway 101
  - Transfers to: **Sprinter** (Oceanside), **Coaster** (Oceanside, Carlsbad Village, Encinitas, Solana Beach), **Amtrak** (Oceanside, Solana Beach), **Metrolink** (Oceanside), **MTS** (La Jolla Shores, Torrey Pines Rd, UTC), **RTA** (Oceanside)
- **302** - Oceanside to Vista Transit Center via Vista Way
  - Transfers to: **Sprinter** (Oceanside, Coast Highway, Vista), **Coaster** (Oceanside), **Amtrak** (Oceanside), **Metrolink** (Oceanside), **RTA** (Oceanside)
- **303** - Oceanside to Vista via Mission Avenue and N Santa Fe Ave
  - Transfers to: **Sprinter** (Oceanside, Vista), **Coaster** (Oceanside), **Amtrak** (Oceanside), **Metrolink** (Oceanside), **RTA** (Oceanside)
- **304*/404*** - Encinitas to San Marcos via Rancho Santa Fe Road
  - Transfers to: **Sprinter** (Palomar College), **Coaster** (Encinitas)
- **305** - Escondido Transit Center to Vista via Mission Rd/S Santa Fe Ave
  - Transfers to: **Sprinter** (Vista, Buena Creek, Palomar College, San Marcos Civic Center, Nordahl Rd, Escondido), **RTA** (Escondido)
- **306** - Fallbrook to Vista via Mission Rd
  - Transfers to: **Sprinter** (Vista)
- **308** - Solana Beach Station to Escondido via Del Dios Hwy.
  - Transfers to: **Sprinter** (Escondido) **Coaster** (Solana Beach), **Amtrak** (Solana Beach), **RTA** (Escondido)
- **309** - Oceanside to Encinitas via El Camino Real
- **313*** - Oceanside to Town Center North via Mesa Drive
  - Transfers to: **Sprinter** (Oceanside), **Coaster** (Oceanside), **Amtrak** (Oceanside), **Metrolink** (Oceanside), **RTA** (Oceanside)
- **315** - Town Center North to 22 Area (Camp Pendleton) via Vandegrift Blvd.
- **317*** - Vandegrift/Gold to Rancho Del Oro Station via Ocean Ranch
  - Transfers to: **Sprinter** (Rancho Del Oro)
- **318**** - Oceanside to El Camino Real Station via Oceanside Blvd
  - Transfers to: **Sprinter** (Oceanside, Coast Highway, Crouch St, El Camino Real), **Coaster** (Oceanside), **Amtrak** (Oceanside), **Metrolink** (Oceanside), **RTA** (Oceanside)
- **319*** - El Corazon Senior Center - Rancho Del Oro Station - MiraCosta College
  - Transfers to: **Sprinter** (Rancho Del Oro)
- **321*** - Carlsbad Village to Palomar College via Legoland
  - Transfers to: **Sprinter** (Palomar College), **Coaster** (Carlsbad Village)
- **325**** - Carlsbad Village to Town Center North via College Blvd

  - Transfers to: **Sprinter** (College Blvd), **Coaster** (Carlsbad Village)
- **331** - Buena Creek Station to University Dr
  - Transfers to: **Sprinter** (Buena Creek)
- **332*** - South Vista and Vista Business Park
  - Transfers to: **Sprinter** (Vista)
- **333** - Town Center North to Vista via Old Grove Rd and Oceanside Blvd
  - Transfers to: **Sprinter** (College Blvd, Melrose Dr, Vista)
- **334*/335*** - Vista Circulator
  - Transfers to: **Sprinter** (Melrose Dr, Vista)
- **340*** - Cal State San Marcos Shuttle
  - Transfers to: **Sprinter** (Cal State San Marcos)
- **347*** - Escondido to Palomar College via Cal State San Marcos
  - Transfers to: **Sprinter** (Palomar College, Cal State San Marcos, Nordahl Rd, Escondido), **RTA** (Escondido)
- **350 Breeze Rapid** - Escondido to Del Lago Park and Ride
  - Transfers to: **Sprinter** (Escondido), **MTS** (Del Lago), **RTA** (Escondido)
- **351/352** - Escondido Circulator
  - Transfers to: **Sprinter** (Escondido), **RTA** (Escondido)
- **354** - Orange Glen High School via Mission Rd, Lincoln Ave, and Midway Dr
  - Transfers to: **Sprinter** (Escondido), **RTA** (Escondido)
- **356** - Morning View Dr, El Norte Pkwy, and Escondido Blvd
  - Transfers to: **Sprinter** (Escondido), **RTA** (Escondido)
- **358*/359*** - N Broadway, Country Club, and El Norte Pkwy
  - Transfers to: **Sprinter** (Escondido), **RTA** (Escondido)
- **386*** - Escondido to Ramona via Wild Animal Park
  - Transfers to: **Sprinter** (Escondido), **MTS** (Ramona), **RTA** (Escondido)
- **388/389** - Escondido to Pala
  - Transfers to: **Sprinter** (Escondido), **RTA** (Escondido)
- **395** - Oceanside to San Clemente via Camp Pendleton
  - Transfers to: **Sprinter** (Oceanside), **Coaster** (Oceanside), **Amtrak** (Oceanside), **Metrolink** (Oceanside), **OCTA** (San Clemente), **RTA** (Oceanside)
- **444*** - Carlsbad Poinsettia - Coaster Connection
  - Transfers to: **Coaster** (Carlsbad Poinsettia)
- **445*** - Carlsbad Poinsettia - Coaster Connection
  - Transfers to: **Coaster** (Carlsbad Poinsettia)

* indicates route that does not run on Saturdays, Sundays, or holidays.

** indicates route that does not run on Sundays or holidays.

Compass Card

Further information: Compass card

In September 2008, SANDAG introduced a new contactless "Compass Card" using technology by Cubic Transportation Systems, Inc. The Compass Card allows passengers from NCTD and the San Diego Metropolitan Transit System to store regional transit passes and/or cash value on a re-writeable RFID card.

Smoking Ban

On October 16, 2008, NCTD board members unanimously voted to adopt NCTD Ordinance 1, banning smoking at all transit centers, train stations, and bus stops. The ban does not apply to parking lots at transit centers and there is no penalty for anyone found noncompliant with the ordinance.[4]

Surfing Madonna

One of the trains ugly overpasses just North of the Encinitas COASTER Station, was beautified by a stunning stain glass and tile mosaic in bluescale of a surfing Virgin Mary in Encinitas, California, called the Surfing Madona, that city is removing it.

No smoking sign at the Vista Transit Center. Notice the butt receptacle placed about 10 feet behind the sign next to the ticket vending machines.

See also

- San Diego Public Transportation

References

[1] (http://www.gonctd.com/pdf/srtp/ch2.pdf)
[2] http://www.signonsandiego.com/news/2009/nov/20/outsourcing-nctds-bus-services-approved/
[3] http://www.gonctd.com/userfiles/file/Board%20Update%2003201l%20-%20040211.pdf
[4] *Minutes of the Special Meeting/Closed Session and the Regular Meeting of North County Transit District* (http://www.gonctd.com/pdf_minutes08/10162008_boardminutes.pdf), 16 Oct 2008, p. 4, , retrieved 6 March 2009

External links

- North County Transit District (http://www.gonctd.com/)
- Transit agency faces $3 million budget gap (http://www.signonsandiego.com/news/northcounty/20080510-9999-1m10transit.html/) (Union Tribune, May 10, 2008)

Transportation_in_San_Diego_County

Transportation in Greater San Diego
Local Rail
San Diego Trolley
COASTER
SPRINTER
Connecting Rail
Metrolink
Pacific Surfliner (Amtrak)
Airports
San Diego International Airport
McClellan-Palomar Airport

The following is a list of transportation in San Diego County, California.

Rail services

Local

San Diego Trolley

The **San Diego Trolley** is a light rail that serves the metropolitan area including central San Diego, East County, South County to the border.

San Diego Coaster

The **San Diego Coast Express Rail**, or **Coaster** is a commuter rail that connects North County to central San Diego. Most riders are commuters who live in North County and work downtown. .

Sprinter

The **Sprinter** is a light rail line between Oceanside and Escondido, California. Fifteen stations are served, including stops in the cities of Oceanside (western terminus), Vista, San Marcos, and Escondido (eastern terminus).

Sprinter service is operated with Siemens Sprinter diesel multiple units manufactured in Germany, where they are widely used.

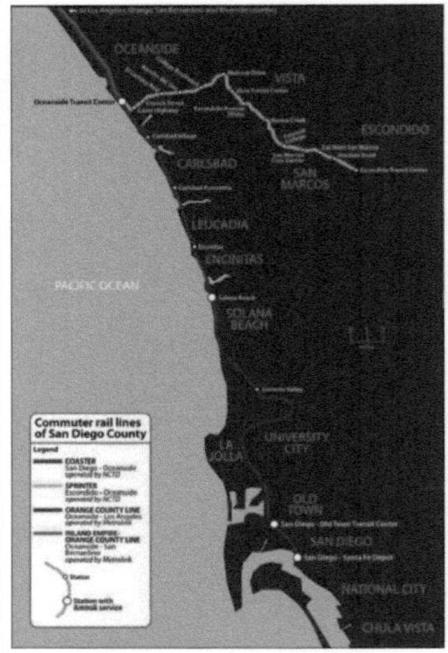

San Diego County commuter rail lines

Linking

Amtrak *Pacific Surfliner*

The ***Pacific Surfliner*** is a 350-mile (563 km) Amtrak passenger train route serving communities on the coast of Southern California between San Diego and San Luis Obispo with stops in Los Angeles.

Metrolink

Metrolink is a commuter rail network connecting major cities in Southern California including San Diego via transfers at Oceanside, Anaheim, Los Angeles, and Riverside. Metrolink does not serve the immediate San Diego area.

Bus services

Buses in San Diego County are operated by two agencies: MTS and NCTD. A third agency, CTS (County Transit System) formerly operated in the eastern and southern suburbs and Poway, but was integrated into MTS in 2002.

Metropolitan Transit System

The San Diego Metropolitan Transit System (MTS) is the public transit service provider for Central, South and East San Diego County. MTS operating subsidiaries include the San Diego Trolley, Incorporated (SDTI) and San Diego Transit, Corporation (SDTC). San Diego Transit directly operates approximately half of all fixed-route bus service within its service area. The remainder are provided under contractual agreements with Veolia Transportation Services or Southland Transportation Services. Some MTS bus routes connect to Breeze services in Escondido or Ramona in the NCTD service area. MTS has begun downloading their route schedules to Google and this has spawned a new generation of trip planners such as TransitTrips [1].

MTS also provides the MTS Access service (provided under contractual agreement with First Transit) throughout its service area. This service is a point-to-point service provided upon request by passengers who qualify for special assistance under the ADA (Americans with Disabilities Act). This service is available in most of the cities of San Diego and Poway, as well as most of the southern and eastern suburbs of San Diego, and connections are available to a similar NCTD service.

The city of Chula Vista contracts with Veolia Transportation Services to provide fixed-route bus service.

North County Transit District

The North County Transit District's (NCTD) "BREEZE" operates 34 bus routes in the northern portion of the county from Del Mar, Escondido, and Ramona, to the Orange County and Riverside County lines.[2] Routes are in the 300 and 400 series (with the exception of route 101).

Bus Passes and transfers may be used both on MTS and NCTD services including the bus and the trolley, but use of the COASTER may require an upgrade.

In addition, express buses from the Riverside Transit Agency connect Temecula with Oceanside.

Road services

Public roads

San Diego County has an extensive network of public roadways that allows vehicle drivers and bicyclists convenient direct access to all practical destinations in the county. Major freeways were built and repeatedly expanded to serve the needs of commuters[3] coming into the city from the suburban regions of North County, South Bay, and East County, as well as the Tijuana metropolitan area.

San Diego is the terminus of three primary interstate highways. Interstate 5 begins at the U.S.-Mexico border and heads north to Los Angeles. Interstate 8 begins at the Sunset Cliffs Boulevard in Ocean Beach, and heads east to the Imperial Valley and Arizona. Interstate 15 begins at Interstate 5 in Southeast San Diego near the 32nd Street Naval Station, then heads northeast to the Inland Empire and then Las Vegas.

Bikeways

Public bikeways in San Diego County consist of Class 1 bike paths, which are separate from the public roads, Class 2 bike lanes, which are demarcated space on the roadways designated to be used primary by bicyclists, and Class 3 bike routes, which are regular surface streets designated as being particularly attractive to bicyclists. SANDAG publishes a map which shows all of these bikeways in the greater San Diego Metropolitan area annually, and it is available at most local bike shops.

Air services

San Diego International Airport is the second busiest single-runway commercial airport in the world, with approximately 600 departures and arrivals carrying 40,000 passengers each day, and over 18 million passengers a year.

However, it occupies a much smaller footprint compared to other airports in the United States. As a result, recently, numerous plans have been proposed to either add another airport in the area, or relocate it entirely.

Ferry services

The Coronado Commuter Ferry takes morning and afternoon commuters across San Diego Bay from the San Diego Broadway Pier, to the Naval Air Station North Island, and the Coronado Ferry Landing [4]. San Diego Harbor Excursion [5] operates ferry shuttles directly to and from the San Diego Broadway Pier, the Coronado Ferry Landing [4] and the Fifth Avenue Landing [6].

References

- San Diego Metropolitan Transit System (MTS) [7]; includes information on MTS Bus and Trolley services
- North San Diego County Transit District [8] - includes information on the Coaster, Sprinter, and Breeze
- San Diego Airport [9]
- Amtrak [10]
- 511 Info [11] - includes information on San Diego Public Transportation, including trip planning
- San Diego Harbor Excursion [5]
- Coronado Ferry Landing [4]

[1] http://transittrips.com
[2] http://www.gonctd.com/breeze/breeze.html
[3] Journal of San Diego History, vol. 41, no. 3, Summer 1995 (http://www.sandiegohistory.org/journal/95summer/chapter21.htm)
[4] http://www.coronadoferrylandingshops.com/
[5] http://www.sdhe.com/san-diego-bay-ferry.html
[6] http://www.fifthavenuelanding.com/
[7] http://www.sdmts.com
[8] http://www.gonctd.com/
[9] http://www.san.org
[10] http://amtrak.com
[11] http://www.transit.511sd.com

North County Transit District

North County Transit District (NCTD)	
NCTD	
Info	
Locale	North San Diego County, California
Transit type	Multi-Modal regional transit network consisting of: Buses Light Rail Commuter rail & Paratransit
Operation	
Began operation	September 30, 1975
Operator(s)	Breeze & Lift: NCTD Sprinter: Veolia Transportation Coaster: TransitAmerica Services
Reporting marks	SDNR
Technical	
Track gauge	4 ft 8½ in (1435 mm) (standard gauge)

The **North County Transit District** (**NCTD**) (reporting mark **SDNR**) is the agency responsible for public transportation in North San Diego County, California, United States. The agency operates the "Breeze" bus service, the Sprinter light rail service between Oceanside and Escondido, and manages the San Diego Coaster, a commuter rail service between Oceanside and Downtown San Diego, in conjunction with Amtrak.

NCTD owns 62 mi (100 km) of mainline track, as well as the 22 mi (35 km) Escondido Branch that is served by the Sprinter beginning in 2008. The Coaster commuter train runs on 41.1 mi (66.1 km) of the mainline.

NCTD maintains two rail yards. One is shared with the San Diego Trolley at 12th & Imperial in Centre City San Diego. It is used to store trainsets during the midday. The other, located north of Oceanside at Stuart Mesa on Camp Pendleton, is shared with Metrolink. This is where the main maintenance facility is located.

History of the San Diego Northern Railway and NCTD

San Diego Northern Railway (SDNR) purchased the tracks used by Coaster from the Atchison, Topeka and Santa Fe Railway in 1994. NCTD created the subsidiary San Diego Northern Railway Corporation in 1994 and dissolved it in 2002.[1] Total investments through NCTD total $750 million.

In November 2009, NCTD approved outsourcing all bus and rail operations effective July 1, 2010 to First Transit. The agency would save $70 million over seven years with the move.[2] Fleet and facility operators remained NCTD employees until their contracts expired June 30, 2011.[3]

NCTD still uses the SDNR reporting marks for its rail equipment. With new rail operations management, between January 2010 and June 2011 ridership increased 29%. This is the single greatest increase in ridership for all US rail-lines in the year 2010.

Services

- **Breeze**- Fixed-route local and express bus services.
- **Breeze Rapid** - Bus rapid transit line in Escondido.
- **Coaster**- Commuter rail service from Oceanside to Downtown San Diego.
- **Lift**- Paratransit service for those with disabilities that are unable to use the accessible fixed-route system.
- **Sprinter**- Light rail from Oceanside to Escondido, California.

Breeze Lines

As of January 2011, there are thirty bus routes run by NCTD.

- **101** - Oceanside Transit Center to UTC via former Highway 101
  - Transfers to: **Sprinter** (Oceanside), **Coaster** (Oceanside, Carlsbad Village, Encinitas, Solana Beach), **Amtrak** (Oceanside, Solana Beach), **Metrolink** (Oceanside), **MTS** (La Jolla Shores, Torrey Pines Rd, UTC), **RTA** (Oceanside)
- **302** - Oceanside to Vista Transit Center via Vista Way
  - Transfers to: **Sprinter** (Oceanside, Coast Highway, Vista), **Coaster** (Oceanside), **Amtrak** (Oceanside), **Metrolink** (Oceanside), **RTA** (Oceanside)
- **303** - Oceanside to Vista via Mission Avenue and N Santa Fe Ave
  - Transfers to: **Sprinter** (Oceanside, Vista), **Coaster** (Oceanside), **Amtrak** (Oceanside), **Metrolink** (Oceanside), **RTA** (Oceanside)
- **304*/404*** - Encinitas to San Marcos via Rancho Santa Fe Road
  - Transfers to: **Sprinter** (Palomar College), **Coaster** (Encinitas)
- **305** - Escondido Transit Center to Vista via Mission Rd/S Santa Fe Ave
  - Transfers to: **Sprinter** (Vista, Buena Creek, Palomar College, San Marcos Civic Center, Nordahl Rd, Escondido), **RTA** (Escondido)
- **306** - Fallbrook to Vista via Mission Rd
  - Transfers to: **Sprinter** (Vista)
- **308** - Solana Beach Station to Escondido via Del Dios Hwy.
  - Transfers to: **Sprinter** (Escondido) **Coaster** (Solana Beach), **Amtrak** (Solana Beach), **RTA** (Escondido)
- **309** - Oceanside to Encinitas via El Camino Real
- **313*** - Oceanside to Town Center North via Mesa Drive
  - Transfers to: **Sprinter** (Oceanside), **Coaster** (Oceanside), **Amtrak** (Oceanside), **Metrolink** (Oceanside), **RTA** (Oceanside)
- **315** - Town Center North to 22 Area (Camp Pendleton) via Vandegrift Blvd.
- **317*** - Vandegrift/Gold to Rancho Del Oro Station via Ocean Ranch
  - Transfers to: **Sprinter** (Rancho Del Oro)
- **318**** - Oceanside to El Camino Real Station via Oceanside Blvd
  - Transfers to: **Sprinter** (Oceanside, Coast Highway, Crouch St, El Camino Real), **Coaster** (Oceanside), **Amtrak** (Oceanside), **Metrolink** (Oceanside), **RTA** (Oceanside)
- **319*** - El Corazon Senior Center - Rancho Del Oro Station - MiraCosta College
  - Transfers to: **Sprinter** (Rancho Del Oro)
- **321*** - Carlsbad Village to Palomar College via Legoland
  - Transfers to: **Sprinter** (Palomar College), **Coaster** (Carlsbad Village)
- **325**** - Carlsbad Village to Town Center North via College Blvd

- Transfers to: **Sprinter** (College Blvd), **Coaster** (Carlsbad Village)
- **331** - Buena Creek Station to University Dr
  - Transfers to: **Sprinter** (Buena Creek)
- **332*** - South Vista and Vista Business Park
  - Transfers to: **Sprinter** (Vista)
- **333** - Town Center North to Vista via Old Grove Rd and Oceanside Blvd
  - Transfers to: **Sprinter** (College Blvd, Melrose Dr, Vista)
- **334*/335*** - Vista Circulator
  - Transfers to: **Sprinter** (Melrose Dr, Vista)
- **340*** - Cal State San Marcos Shuttle
  - Transfers to: **Sprinter** (Cal State San Marcos)
- **347*** - Escondido to Palomar College via Cal State San Marcos
  - Transfers to: **Sprinter** (Palomar College, Cal State San Marcos, Nordahl Rd, Escondido), **RTA** (Escondido)
- **350 Breeze Rapid** - Escondido to Del Lago Park and Ride
  - Transfers to: **Sprinter** (Escondido), **MTS** (Del Lago), **RTA** (Escondido)
- **351/352** - Escondido Circulator
  - Transfers to: **Sprinter** (Escondido), **RTA** (Escondido)
- **354** - Orange Glen High School via Mission Rd, Lincoln Ave, and Midway Dr
  - Transfers to: **Sprinter** (Escondido), **RTA** (Escondido)
- **356** - Morning View Dr, El Norte Pkwy, and Escondido Blvd
  - Transfers to: **Sprinter** (Escondido), **RTA** (Escondido)
- **358*/359*** - N Broadway, Country Club, and El Norte Pkwy
  - Transfers to: **Sprinter** (Escondido), **RTA** (Escondido)
- **386*** - Escondido to Ramona via Wild Animal Park
  - Transfers to: **Sprinter** (Escondido), **MTS** (Ramona), **RTA** (Escondido)
- **388/389** - Escondido to Pala
  - Transfers to: **Sprinter** (Escondido), **RTA** (Escondido)
- **395** - Oceanside to San Clemente via Camp Pendleton
  - Transfers to: **Sprinter** (Oceanside), **Coaster** (Oceanside), **Amtrak** (Oceanside), **Metrolink** (Oceanside), **OCTA** (San Clemente), **RTA** (Oceanside)
- **444*** - Carlsbad Poinsettia - Coaster Connection
  - Transfers to: **Coaster** (Carlsbad Poinsettia)
- **445*** - Carlsbad Poinsettia - Coaster Connection
  - Transfers to: **Coaster** (Carlsbad Poinsettia)

***** indicates route that does not run on Saturdays, Sundays, or holidays.

****** indicates route that does not run on Sundays or holidays.

Compass Card

Further information: Compass card

In September 2008, SANDAG introduced a new contactless "Compass Card" using technology by Cubic Transportation Systems, Inc. The Compass Card allows passengers from NCTD and the San Diego Metropolitan Transit System to store regional transit passes and/or cash value on a re-writeable RFID card.

Smoking Ban

On October 16, 2008, NCTD board members unanimously voted to adopt NCTD Ordinance 1, banning smoking at all transit centers, train stations, and bus stops. The ban does not apply to parking lots at transit centers and there is no penalty for anyone found noncompliant with the ordinance.[4]

Surfing Madonna

One of the trains ugly overpasses just North of the Encinitas COASTER Station, was beautified by a stunning stain glass and tile mosaic in bluescale of a surfing Virgin Mary in Encinitas, California, called the Surfing Madona, that city is removing it.

No smoking sign at the Vista Transit Center. Notice the butt receptacle placed about 10 feet behind the sign next to the ticket vending machines.

See also

- San Diego Public Transportation

References

[1] (http://www.gonctd.com/pdf/srtp/ch2.pdf)

[2] http://www.signonsandiego.com/news/2009/nov/20/outsourcing-nctds-bus-services-approved/

[3] http://www.gonctd.com/userfiles/file/Board%20Update%2003201l%20-%2004021l.pdf

[4] *Minutes of the Special Meeting/Closed Session and the Regular Meeting of North County Transit District* (http://www.gonctd.com/pdf_minutes08/10162008_boardminutes.pdf), 16 Oct 2008, p. 4, , retrieved 6 March 2009

External links

- North County Transit District (http://www.gonctd.com/)
- Transit agency faces $3 million budget gap (http://www.signonsandiego.com/news/northcounty/20080510-9999-1m10transit.html/) (Union Tribune, May 10, 2008)

Article Sources and Contributors

Sprinter_(North_County_Transit_District) *Source*: http://en.wikipedia.org/w/index.php?title=Sprinter_%28North_County_Transit_District%29 *Contributors*: 21655, AkiStuart, Arsenikk, Auntof6, Beagel, BlkMtn, Branddobbe, Butros, Can't Undo, Ccarmichael, Cnbrb, Cooljuno411, DavDaven, DavidArthur, DavidCiani, Ebear422, Evil saltine, GNRY09, Hawaiian717, Jfruh, John of Reading, Koavf, Koman90, Kylephoto760, Mackensen, MarsRover, Metroman2, Mikus, Millbrooky, Moebiusuibeom-en, MrHudson, Mschmidt62, Mulad, NE2, Njbob, Nutmegger, Oakshade, Oddperson12, OettingerCroat, Oknazevad, Optigan13, Parent5446, Pitamakan, Plasma east, Plattypus1, RickyCourtney, Rlcoleman, Salcan, Sameboat, SameerKhan, Savewales, TexasAndroid, The Port of Authority, Torritorri, Traal, VNPeO, Vasilken, VitaleBaby, Vrysxy, Wickifrank, Wlindley, Yellowdesk, Zscout370, Zyxw, 57 anonymous edits

Diesel_locomotive *Source*: http://en.wikipedia.org/w/index.php?title=Diesel_locomotive *Contributors*: 37427, 7severn7, AMCKen, AVRS, Aaronawernick, Aitias, Alco83, Anatoly.bourov, Andre Kritzinger, Andres, Angryxpeh, Anthony Ivanoff, Ardric47, Attilios, Axpde, Beano, Bigdumbdinosaur, Biscuittin, Bkil, Brimacl, Cancun771, Capricorn42, Carnildo, Casey56, CesarB, Chris j wood, Coffeezombie, Coolhawks88, DB 103245, DMChatterton, Davewho2, David Jordan, Dellarb, DocWatson42, Dp67, Droob, EAderhold, EdJogg, Edward, Elkman, Fastilysock, Favonian, Felipe P, FengRail, Fernando1234, Ferritecore, Fyrael, Geschichte, GoingBatty, Gwernol, Harfarhs, HarveyHenkelmann, Hellbus, Highspeed, Hmains, Hrsn, Hugo999, Ilaiho, Iridescent, Ironmani, JGHowes, Jackehammond, Jagan123, Jameslwoodward, Jauhienij, Jklamo, Jonny55022, JzoJames, KJRehberg, Kelisi, Koavf, Kyteto, LMB, Leader Hosing, Lensovet, Lightmouse, LordJesseD, LorenzoB, Mac, MakeChooChooGoNow, Mangoe, Meggar, Michael Daly, MickMacNee, Mikus, Mindmatrix, Mopza, Morven, Mzmadmike, NE2, Natalie Erin, NerdyScienceDude, Nick Number, Nickg1980, Nictrain, Nikola Smolenski, Noleander, Nopetro, Peter Horn, Piledhigheranddeeper, R'n'B, Railmodel, Rjwilmsi, Rks13, Robbie aka Zoqaeski, Rogermw, RoySmith, Rpdant767, Rrostrom, S3000, Saval, SayCheeeeeese, SchuminWeb, Sheeju, ShelfSkewed, Soc8675309, SpaceChimp1992, SriMesh, Suckindiesel, Sulzer55, Synthetik, TAG.Odessa, Tabletop, Tewfik, Tim PF, Timtrent, Tmollerup, Tohd8BohaithuGh1, TonyTheTiger, Treekids, Ummit, Useddenim, Vegaswikian, Vytautas LT, Wdl1961, Wongm, Woohookitty, Wtshymanski, Wuhwuzdat, Wwheaton, XXV, Yoshirocks8, , 149 anonymous edits

Oceanside,_California *Source*: http://en.wikipedia.org/w/index.php?title=Oceanside%2C_California *Contributors*: 08OceanBeach SD, 1958publius, 72Dino, Abby, Acalamari, Akendall, Alish128, AliveFreeHappy, Andrewpmk, AndyAldrin, AngChenrui, Antandrus, AresAndEnyo, Ashnbham, Axion22, Baal32, Bart133, Bellhalla, Ben Lunsford, Bender235, Berwil 92083, BlankVerse, Bongwarrior, Bovineboy2008, Brakspear, Branddobbe, BrandonFurtado, Bryan Derksen, CJLL Wright, CL, CaliE75, Camerong, Candyman 53, Carlossuarez46, Chris the speller, ChrisCork, Christian75, Cifchamp0809, Classic49er, Clementina, Clintville, Cooljuno411, Cupoladeesco, Cyanidethistles, Damicatz, Danbrown415, Davemcarlson, Daystar1964, Denisarona, DirkFunk, Downwards, Drewfarr, Dual Freq, DylanW, Eco84, Edgeshappy12, Endo-Plexor, Enigmaman, Etcher, FX 72.197.133.100, Fabrictramp, Fæ, Gadfium, GalaazV, Gellersen, Genghiskhanviet, Geoking66, Gogo Dodo, Golbez, Graeme Bartlett, Greteck, Gunmetal Angel, Hankw78, Happyfatman021, Harland1, Hektor, Hmains, House1090, Hu12, Ironsharp, JamesBWatson, Jamjam0621, Jetstone, John K, John254, Jojhutton, JonHarder, Juliancolton, Jvcdude, K1Bond007, Kbdank71, Kevin W., Kmandigm, Kusunose, Kyle1278, L Kensington, Largoplazo, Lightmouse, Lobsterman5468, Look2See1, MJCdetroit, Magichands, Manhoor69, Massens, Maurreen, Medmura118, Michael Patrick, Mike D 26, Mike Rosoft, Mm40, Mobussell, Monkey Bounce, Morgankevinj huggle, Mrmuk, NorwegianBlue, Ntsimp, Ohnoitsjamie, Optigan13, Orangemike, Packerfansam, Persian Poet Gal, Phildgeo, Philip Trueman, President Rhapsody, Pretendo, Professorgupta, Ptsullivan79, Qrsdogg, RFD, Ram-Man, Raynardo, Rich Farmbrough, RightCowLeftCoast, Rnickel, RodrigoStarlight, Rosiestep, Ryuhaku, SQL, Sango123, Saxbryn, Scprior2000, Sctechlaw, Seashinegirl, Sedna10387, Seven Days, Sgt Simpson, Shereth, Short Verses, Shortride, Silas101, SoCal L.A., Socal gal at heart, Some jerk on the Internet, SpikeTorontoRCP, Stamford99, Stefan da, Stevey7788, Surfingcal, Syphonbyte, Tassedethe, Tempodivalse, Tenebrae, That Guy, From That Show!, The Elephants, The wub, Thecheesykid, Themaindude, Tmcmu604, TonyTheTiger, Torritorri, Trusilver, Ulric1313, Uriel8, Vanka5, Vrysxy, Warrush, Wawzat, Whhalbert, WhisperToMe, White Lightning, Wikibofh, Will Beback, Woohookitty, WookieInHeat, Worldenc, Yair rand, Zaxem, Zzyzx11, 371 anonymous edits

Escondido,_California *Source*: http://en.wikipedia.org/w/index.php?title=Escondido%2C_California *Contributors*: 08OceanBeach SD, 1958publius, 4117Ashford, Acalamari, Acatjr55, Adagio for Strings, Agne27, Agüeybaná, Alansohn, AliveFreeHappy, Arch dude, Berwil 92083, Bibliomaniac15, Biederman, Biruitorul, Bjpcolo, BlankVerse, Bobby H. Heffley, Branddobbe, Bryan Derksen, Bwbullough, Calvin 1998, Carlossuarez46, CharlotteWebb, Chowbok, Chris Doig, ChrisCork, Chuglur, Clockwork Laser, Colinrrmf, Cooljuno411, Cotoco, Crumb, Cubelodyte, DARTH SIDIOUS 2, Daniel Christensen, Dario D., Dark Charles, Darklilac, Darksome, Dcmcgov, Deflective, Delgstewart, Diwas, Djinn112, Dlifedesign, Dogru144, Drumnstas, Edward, Elockid, Emargie, Epbr123, Esclib, EskoCruz, Esprqii, Etcher, Evil saltine, Excirial, Franklin Dmitryev, Frisch1, GTBacchus, Gadfium, Galaxiaad, Gellersen, Gilliam, Giraffedata, Gouryella, GreenGlass1972, Gregorknight, Greteck, Gyrferret, HOLLYROAST1, Hmains, House1090, Hu12, Indoorsoccer, Itinerant1, Jack Krauser, John K, Johnpacklambert, Jojhutton, K1Bond007, KC0ZHQ, Katieragazzi, Kristaw, Kwamikagami, Kylephoto760, Lateg, Lightmouse, Look2See1, Lotje, MJCdetroit, Marilee67, Mario Žamić, Mayur, Mcanmoocanu, MelanieN, Mike D 26, Mscuthbert, Mtthomps89, Mytrueeditor, NE2, NapoliRoma, Nehrams2020, Neumannk, Next Fiday, Night of the Big Wind, Node ue, Norbu, Ohnoitsjamie, Oknazevad, Optigan13, Pandelver, Parsa, Peter Horn, Phildgeo, Pristino, Puchiko, R32GTR, RA0808, Ram-Man, Rdunbariii, Rich Farmbrough, Rigadoun, RightCowLeftCoast, Rodan44, Ryuhaku, Scamper, Scott5114, Secret (renamed), Sedna10387, Seven Days, Shereth, Short Verses, Shortride, SimMoonXP, Smalljim, Smeira, SoCal L.A., SoCalDonF, Socal gal at heart, Socalglamis, Softweyr, Soltras, StarTrekkie, Stevejess, Stevey7788, Sum1uallno, Talatra, Tamfang, Tesscass, The Elephants, Thenexxuz, Tide rolls, Timeineurope, Tktru, Tommybender, TonyTheTiger, Trackway, Trevtroy, Trusilver, Ufossuck, Ulric1313, User2004, Ux dave, Uyvsdi, Vanka5, WBarker, Wasbeer, Well-Read Red, Whhalbert, WhisperToMe, WikHead, Wikialertoregon, Will Beback, Woohookitty, Worldenc, Wtmitchell, X65, Xenocidic, Yamidointhis, Yonskii, Zuzbie8, Zzyzx11, 279 anonymous edits

Coaster_(San_Diego) *Source*: http://en.wikipedia.org/w/index.php?title=Coaster_%28San_Diego%29 *Contributors*: 08OceanBeach SD, 123456789gohgfohngo, AkiStuart, AmtrakSurfliner768, Anomalocaris, Anwyn19, Arsenikk, BD2412, BariVaz, Bishop^, Branddobbe, Bunthorne, Butros, C33, Camrock, Ccarmichael, Chris the speller, ChrisRuvolo, CoasterMike2105, Colonies Chris, Cooljuno411, Corgi596, DanTD, DavDaven, Djackino, Dohn joe, Drbreznjev, Evanparker, Fairlyoddparents1234, Ghoffman, Glenn L, Hawaiian717, Ian Pitchford, Iliketimmyturner, It's me Michael, JYolkowski, Jaraalbe, JoseAwfulman, JulieADriver, Kbdank71, Kenyon, Koavf, Koman90, Kylephoto760, Levineps, Lost on belmont, Mackensen, Marriedtofilm, MasterA113, MelanieN, Mercurywoodrose, Michaellovestrains, MrHudson, Nebula2357, Neilc, Niteowlneils, Ntsimp, Nutmegger, Ojw, Oknazevad, Ospalh, PSurfliner, PerryPlanet, Plasma east, PoccilScript, Poem, ProjectSavior, Qnonsense, Redtroll, Rhallanger, RickyCourtney, SPUI, Salcan, Sameboat, SameerKhan, Shortride, Smedpull, Stevan.rich, Synthesiz, Tech Teacher 101, TexasAndroid, The Port of Authority, The06bug, Train2104, Ulric1313, User2004, Vrysxy, WadeSimMiser, Wlindley, Zian, Zyxw, Zzyzx11, 76 anonymous edits

Palomar_College *Source*: http://en.wikipedia.org/w/index.php?title=Palomar_College *Contributors*: 4117Ashford, AP1787, Aelfthrytha, AkiStuart, Azumanga1, BD2412, Beginning, Branddobbe, Clnitely, D6, DGG, Esprqii, Evil saltine, Johnpacklambert, Kamezuki, Kenyon, Kingdomkid90, Larbox, Masonpatriot, Mcanmoocanu, Mgiganteus1, Mjolnir1134, NerdyScienceDude, OccamzRazor, Peter Horn, Rich Farmbrough, Roxy34, Sam Blacketer, Samxli, Shimgray, SimMoonXP, Skier Dude, Snespeca, Stepheng3, StrangeWill, TRBP, TheParanoidOne, Therealjax, ValleyWikian, VectorPosse, Wikialertoregon, ZiggySD, 58 anonymous edits

California_State_University,_San_Marcos *Source*: http://en.wikipedia.org/w/index.php?title=California_State_University%2C_San_Marcos *Contributors*: 72Dino, Adam McMaster, AkiStuart, Americanadian, Amerique, Angela, Anon134, Anon92096, Aqueous07, Arate001, Asiananimal, Aspects, Baysunghur@yahoo.com, Bobblewik, Bobo192, CaliEd, Catapult, Cerebralpayne, Coldmachine, Cpastern, D6, DO'Neil, Dabackgammonator, Dbwarren, Deli nk, Discospinster, Dynaflow, Eamuscatuli, ElKevbo, Eliz81, Esrever, Evil saltine, Fpo, Gadfium, Gene Nygaard, Gentgeen, J3ff, Jaraalbe, Jigen III, Joelwest, Kevin Ryde, Kinyanni, Lightmouse, M.Fanavar, Marco Guzman, Jr, Masonpatriot, Mithrandir, Niteowlneils, Oplm da god, Ortcutt, Peter Horn, Phoebe, RJN, Ritual, Rjwoer, Rocksanddirt, Rosiestep, Rtdrury, SanMarcos, Serogi, Seven Days, Siva1979, Stealthound, Stepheng3, Streltzer, Tassedethe, Tomlillis, Usgnus, ValleyWikian, Zzyzx11 the clone, 149 anonymous edits

North_County,_San_Diego *Source*: http://en.wikipedia.org/w/index.php?title=North_County%2C_San_Diego *Contributors*: 08OceanBeach SD, 72Dino, AThing, BlankVerse, Branddobbe, BronzeAddy, Cooljuno411, Dave Andrew, Dohn joe, Evil saltine, Fratrep, Gill Giller Gillerger, House1090, J04n, Jllm06, Joelwest, Kylephoto760, Look2See1, MalcolmHeardSB, NE2, Neilc, Rick Block, Rnickel, Rschen7754, SchutteGod, Ski09, SoCal L.A., Techieman, WhisperToMe, XR9, Ylee, Zzyzx11, 24 anonymous edits

North_County_Transit_District *Source*: http://en.wikipedia.org/w/index.php?title=North_County_Transit_District *Contributors*: Arsenikk, Azumanga1, Blaaaa324345, Branddobbe, Calwatch, Cbrek04, Cooljuno411, CyLWatkins, DanTD, Evil saltine, Fastily, Hawaiian717, JoseAwfulman, Jrg7787, Koman90, Michaellovestrains, Narayansg, Plasma east, Poem, RickyCourtney, Robert Moore, SameerKhan, Scott Sanchez, Secondarywaltz, Smedpull, The Port of Authority, Track Legs, Vrysxy, Yellowdesk, 10 anonymous edits

Transportation_in_San_Diego_County *Source*: http://en.wikipedia.org/w/index.php?title=Transportation_in_San_Diego_County *Contributors*: (aeropagitica), Auntof6, Basawala, Born2cycle, Branddobbe, Calwatch, Can't Undo, Cooljuno411, Dashira, Dmbodinger, Erechtheus, Evil saltine, Hawaiian717, Look2See1, MelanieN, Pcapozzi, Plasma east, Pzoxicuvybtnrm, Rfc1394, Rich Farmbrough, SameerKhan, SoCalSuperEagle, The Port of Authority, Tinlinkin, Track Legs, Ulric1313, Vrysxy, WOSlinker, WhisperToMe, Woohookitty, XinJeisan, YUL89YYZ, Zian, Zyxw, Zzyzx11, 32 anonymous edits

North County Transit District *Source*: http://en.wikipedia.org/w/index.php?title=North_County_Transit_District *Contributors*: Arsenikk, Azumanga1, Blaaaa324345, Branddobbe, Calwatch, Cbrek04, Cooljuno411, CyLWatkins, DanTD, Evil saltine, Fastily, Hawaiian717, JoseAwfulman, Jrg7787, Koman90, Michaellovestrains, Narayansg, Plasma east, Poem, RickyCourtney, Robert Moore, SameerKhan, Scott Sanchez, Secondarywaltz, Smedpull, The Port of Authority, Track Legs, Vrysxy, Yellowdesk, 10 anonymous edits

Image Sources, Licenses and Contributors

file:SPRINTER logo.svg *Source*: http://en.wikipedia.org/w/index.php?title=File:SPRINTER_logo.svg *License*: unknown *Contributors*: Koman90, Skier Dude, 3 anonymous edits

Image:San Diego commuter rail map.png *Source*: http://en.wikipedia.org/w/index.php?title=File:San_Diego_commuter_rail_map.png *License*: unknown *Contributors*: The Port of Authority (talk). Original uploader was The Port of Authority at en.wikipedia

File:Sprinter at Oceanside.jpg *Source*: http://en.wikipedia.org/w/index.php?title=File:Sprinter_at_Oceanside.jpg *License*: unknown *Contributors*: William Lindley

File:Sprinter light rail train.jpg *Source*: http://en.wikipedia.org/w/index.php?title=File:Sprinter_light_rail_train.jpg *License*: unknown *Contributors*: Mikus (talk). Original uploader was Mikus at en.wikipedia

Image:Three-loco-styles.jpg *Source*: http://en.wikipedia.org/w/index.php?title=File:Three-loco-styles.jpg *License*: unknown *Contributors*: User:Wongm

Image:Dawlish Warren 1970s - 5.jpg *Source*: http://en.wikipedia.org/w/index.php?title=File:Dawlish_Warren_1970s_-_5.jpg *License*: unknown *Contributors*: Mike Powell

Image:2M62M-1198 Kyviskes-Kena.jpg *Source*: http://en.wikipedia.org/w/index.php?title=File:2M62M-1198_Kyviskes-Kena.jpg *License*: unknown *Contributors*: User:Vytautas LT

Image:Indian Diesel Loco.JPG *Source*: http://en.wikipedia.org/w/index.php?title=File:Indian_Diesel_Loco.JPG *License*: unknown *Contributors*: User:Kumar83

Image:D2069 at Doncaster Works.JPG *Source*: http://en.wikipedia.org/w/index.php?title=File:D2069_at_Doncaster_Works.JPG *License*: unknown *Contributors*: Phil Scott (w:en:User:Our PhellapUser:Our Phellap)

File:DieselMechanicalLocomotiveSchematic.svg *Source*: http://en.wikipedia.org/w/index.php?title=File:DieselMechanicalLocomotiveSchematic.svg *License*: unknown *Contributors*: User:Noleander

File:DieselElectricLocomotiveSchematic.svg *Source*: http://en.wikipedia.org/w/index.php?title=File:DieselElectricLocomotiveSchematic.svg *License*: unknown *Contributors*: User:Noleander

File:Metra Locomotives F40PH-2 & MP36PH-3S.jpg *Source*: http://en.wikipedia.org/w/index.php?title=File:Metra_Locomotives_F40PH-2_&_MP36PH-3S.jpg *License*: unknown *Contributors*: User:Jovianeye

Image:2TE10U Russian Locomotive.jpg *Source*: http://en.wikipedia.org/w/index.php?title=File:2TE10U_Russian_Locomotive.jpg *License*: unknown *Contributors*: User:Anthony Ivanoff

Image:CSX locomotive cab.jpg *Source*: http://en.wikipedia.org/w/index.php?title=File:CSX_locomotive_cab.jpg *License*: unknown *Contributors*: User:JGHowes

Image:S3DieselElectricLocomotiveSRM.jpg *Source*: http://en.wikipedia.org/w/index.php?title=File:S3DieselElectricLocomotiveSRM.jpg *License*: unknown *Contributors*: User:SriMesh

Image:12567B.jpg *Source*: http://en.wikipedia.org/w/index.php?title=File:12567B.jpg *License*: unknown *Contributors*: Original uploader was Chesapeake at en.wikipedia

Image:3000hp curve ver2.jpg *Source*: http://en.wikipedia.org/w/index.php?title=File:3000hp_curve_ver2.jpg *License*: unknown *Contributors*: Original uploader was Suckindiesel at en.wikipedia

Image:GE Genesis P32AC-DM 202 MNCRR Ossining.jpg *Source*: http://en.wikipedia.org/w/index.php?title=File:GE_Genesis_P32AC-DM_202_MNCRR_Ossining.jpg *License*: unknown *Contributors*: User:bebo2good1. Original uploader was Bebo2good1 at en.wikipedia

Image:Db-220002-01-c.jpg *Source*: http://en.wikipedia.org/w/index.php?title=File:Db-220002-01-c.jpg *License*: unknown *Contributors*: User:Kjunix

Image:Western Warship and Hymek.jpg *Source*: http://en.wikipedia.org/w/index.php?title=File:Western_Warship_and_Hymek.jpg *License*: unknown *Contributors*: Les Chatfield

Image:VR Dv12 locomotive in Tampere Aug2008 001.jpg *Source*: http://en.wikipedia.org/w/index.php?title=File:VR_Dv12_locomotive_in_Tampere_Aug2008_001.jpg *License*: unknown *Contributors*: User:Methem

Image:DH GMD 6031 VFRGS-RF Guido Mota.jpg *Source*: http://en.wikipedia.org/w/index.php?title=File:DH_GMD_6031_VFRGS-RF_Guido_Mota.jpg *License*: unknown *Contributors*: Imagem publicada na Revista Ferroviária. Cortesia: Alfredo Rodrigues

Image:GM-EMD 2000.jpg *Source*: http://en.wikipedia.org/w/index.php?title=File:GM-EMD_2000.jpg *License*: unknown *Contributors*: Daniel Sichelschmidt

File:Canadian National diesels southbound on the Norfolk Southern near Polaris Mall in Columbus, Ohio (2092116881).jpg *Source*: http://en.wikipedia.org/w/index.php?title=File:Canadian_National_diesels_southbound_on_the_Norfolk_Southern_near_Polaris_Mall_in_Columbus,_Ohio_(2092116881).jpg *License*: unknown *Contributors*: geocam20000

File:Oceanside Beach Panorama.jpg *Source*: http://en.wikipedia.org/w/index.php?title=File:Oceanside_Beach_Panorama.jpg *License*: unknown *Contributors*: Randy OCH

File:San_Diego_County_California_Incorporated_and_Unincorporated_areas_Oceanside_Highlighted.svg *Source*: http://en.wikipedia.org/w/index.php?title=File:San_Diego_County_California_Incorporated_and_Unincorporated_areas_Oceanside_Highlighted.svg *License*: unknown *Contributors*: Arkyan

file:Usa edcp location map.svg *Source*: http://en.wikipedia.org/w/index.php?title=File:Usa_edcp_location_map.svg *License*: unknown *Contributors*: User:Uwe Dedering

File:Red pog.svg *Source*: http://en.wikipedia.org/w/index.php?title=File:Red_pog.svg *License*: unknown *Contributors*: Anomie

File:Flag of the United States.svg *Source*: http://en.wikipedia.org/w/index.php?title=File:Flag_of_the_United_States.svg *License*: unknown *Contributors*: Anomie

File:Flag of California.svg *Source*: http://en.wikipedia.org/w/index.php?title=File:Flag_of_California.svg *License*: unknown *Contributors*: w:en:User:DevinCookDevin Cook

File:Andrew Jackson Myers.jpg *Source*: http://en.wikipedia.org/w/index.php?title=File:Andrew_Jackson_Myers.jpg *License*: unknown *Contributors*: Jot Powers

File:Oceanside city hall.jpg *Source*: http://en.wikipedia.org/w/index.php?title=File:Oceanside_city_hall.jpg *License*: unknown *Contributors*: Jot Powers

File:Oside blvd.png *Source*: http://en.wikipedia.org/w/index.php?title=File:Oside_blvd.png *License*: unknown *Contributors*: User:Drewfarr

File:Oceanside, CA Marina.jpg *Source*: http://en.wikipedia.org/w/index.php?title=File:Oceanside,_CA_Marina.jpg *License*: unknown *Contributors*: surfers

File:Oceanside Pier (2).JPG *Source*: http://en.wikipedia.org/w/index.php?title=File:Oceanside_Pier_(2).JPG *License*: unknown *Contributors*: User:Cupoladeesco

File:Oceanside Pier.jpg *Source*: http://en.wikipedia.org/w/index.php?title=File:Oceanside_Pier.jpg *License*: unknown *Contributors*: Bill Newman

File:Flag of Mexico.svg *Source*: http://en.wikipedia.org/w/index.php?title=File:Flag_of_Mexico.svg *License*: unknown *Contributors*: User:AlexCovarrubias

File:Flag of Japan.svg *Source*: http://en.wikipedia.org/w/index.php?title=File:Flag_of_Japan.svg *License*: unknown *Contributors*: Anomie

File:Flag of Samoa.svg *Source*: http://en.wikipedia.org/w/index.php?title=File:Flag_of_Samoa.svg *License*: unknown *Contributors*: Alkari, Anime Addict AA, Avala, Emijrp, Fry1989, Homo lupus, Klemen Kocjancic, Kookaburra, Kronocide, Mattes, Nightstallion, TFCforever, Telim tor, Vzb83, Xufanc, Zscout370, 1 anonymous edits

File:Grand01.jpg *Source*: http://en.wikipedia.org/w/index.php?title=File:Grand01.jpg *License*: unknown *Contributors*: User:Dcmcgov

File:San_Diego_County_California_Incorporated_and_Unincorporated_areas_Escondido_Highlighted.svg *Source*: http://en.wikipedia.org/w/index.php?title=File:San_Diego_County_California_Incorporated_and_Unincorporated_areas_Escondido_Highlighted.svg *License*: unknown *Contributors*: Arkyan

File:Loudspeaker.svg *Source*: http://en.wikipedia.org/w/index.php?title=File:Loudspeaker.svg *License*: unknown *Contributors*: Bayo, Gmaxwell, Husky, Iamunknown, Mirithing, Myself488, Nethac DIU, Omegatron, Rocket000, The Evil IP address, Wouterhagens, 18 anonymous edits

Image:Looking south west across Escondido .jpg *Source*: http://en.wikipedia.org/w/index.php?title=File:Looking_south_west_across_Escondido_.jpg *License*: unknown *Contributors*: Chris Doig

Image:Grand Ave Escondido.jpg *Source*: http://en.wikipedia.org/w/index.php?title=File:Grand_Ave_Escondido.jpg *License*: unknown *Contributors*: Chris Doig

Image:Grape Day Park Escondido.jpg *Source*: http://en.wikipedia.org/w/index.php?title=File:Grape_Day_Park_Escondido.jpg *License*: unknown *Contributors*: Chris Doig

Image:Lake Dixon Escondido CA.jpg *Source*: http://en.wikipedia.org/w/index.php?title=File:Lake_Dixon_Escondido_CA.jpg *License*: unknown *Contributors*: Chris Doig

Image:Escondido City Hall.jpg *Source*: http://en.wikipedia.org/w/index.php?title=File:Escondido_City_Hall.jpg *License*: unknown *Contributors*: Chris Doig

file:COASTER_logo.svg *Source*: http://en.wikipedia.org/w/index.php?title=File:COASTER_logo.svg *License*: unknown *Contributors*: User:SreeBot

Image:coaster2.jpg *Source*: http://en.wikipedia.org/w/index.php?title=File:Coaster2.jpg *License*: unknown *Contributors*: Original uploader was MrHudson at en.wikipedia

Image:SOL Amtrak Station.jpg *Source*: http://en.wikipedia.org/w/index.php?title=File:SOL_Amtrak_Station.jpg *License*: unknown *Contributors*: User:Jorobeq. Original uploader was Jorobeq at en.wikipedia

Image:Coaster F40PHM-2C 2104 at Solana Beach, CA.JPG *Source*: http://en.wikipedia.org/w/index.php?title=File:Coaster_F40PHM-2C_2104_at_Solana_Beach,_CA.JPG *License*: unknown *Contributors*: User:AmtrakSurfliner768

Image:Palomar-college-logo.png *Source*: http://en.wikipedia.org/w/index.php?title=File:Palomar-college-logo.png *License*: unknown *Contributors*: Calmer Waters, Kenyon, Nehrams2020, Sfan00 IMG, Tom Morris, 1 anonymous edits

File:CSU_San_Marcos_Seal.jpg *Source*: http://en.wikipedia.org/w/index.php?title=File:CSU_San_Marcos_Seal.jpg *License*: unknown *Contributors*: Jigen III

File:CSU San Marcos 20th Anniversary.png *Source*: http://en.wikipedia.org/w/index.php?title=File:CSU_San_Marcos_20th_Anniversary.png *License*: unknown *Contributors*: Ritual, Sfan00 IMG

File:Oceansidepier.jpg *Source*: http://en.wikipedia.org/w/index.php?title=File:Oceansidepier.jpg *License*: unknown *Contributors*: User:Alessandriana

File:North County San Diego.png *Source*: http://en.wikipedia.org/w/index.php?title=File:North_County_San_Diego.png *License*: unknown *Contributors*: User:Cooljuno411

File:Banded juvenile California Least Tern.jpg *Source*: http://en.wikipedia.org/w/index.php?title=File:Banded_juvenile_California_Least_Tern.jpg *License*: unknown *Contributors*: Linda Tanner

Image:Del-Mar-Race-Track.jpg *Source*: http://en.wikipedia.org/w/index.php?title=File:Del-Mar-Race-Track.jpg *License*: unknown *Contributors*: Helenaak, Intersofia, Mrrxx, NapoliRoma, Stan Shebs

Image:Oceanside Beach Panorama.jpg *Source*: http://en.wikipedia.org/w/index.php?title=File:Oceanside_Beach_Panorama.jpg *License*: unknown *Contributors*: Randy OCH

Image:Grand01.jpg *Source*: http://en.wikipedia.org/w/index.php?title=File:Grand01.jpg *License*: unknown *Contributors*: User:Dcmcgov

Image:DMTC july 2008.jpg *Source*: http://en.wikipedia.org/w/index.php?title=File:DMTC_july_2008.jpg *License*: unknown *Contributors*: User:Achen33

Image:Lakesanmarcos.jpg *Source*: http://en.wikipedia.org/w/index.php?title=File:Lakesanmarcos.jpg *License*: unknown *Contributors*: User:Taylorj661

Image:Oceanside Pier.jpg *Source*: http://en.wikipedia.org/w/index.php?title=File:Oceanside_Pier.jpg *License*: unknown *Contributors*: Bill Newman

Image:TwinPeaksPoway1.jpg *Source*: http://en.wikipedia.org/w/index.php?title=File:TwinPeaksPoway1.jpg *License*: unknown *Contributors*: Perdelsky

File:NCTD.svg *Source*: http://en.wikipedia.org/w/index.php?title=File:NCTD.svg *License*: unknown *Contributors*: User:BetacommandBot, User:CmdrObot, User:Cydebot, User:ESkog, User:Polbot, User:RickyCourtney

Image:Vista Transit 12-26-08.jpg *Source*: http://en.wikipedia.org/w/index.php?title=File:Vista_Transit_12-26-08.jpg *License*: unknown *Contributors*: User:Smedpull

Image:SD Trolley Green Line.jpg *Source*: http://en.wikipedia.org/w/index.php?title=File:SD_Trolley_Green_Line.jpg *License*: unknown *Contributors*: Thomas Baer. Original uploader was Track Legs at en.wikipedia

MIX
Papier aus verantwortungsvollen Quellen
Paper from responsible sources
FSC® C105338

Printed by Books on Demand GmbH, Norderstedt / Germany